GETTING STARTED WITH STATA
FOR WINDOWS®

A Stata Press Publication
StataCorp LP
College Station, Texas

Stata Press, 4905 Lakeway Drive, College Station, Texas 77845

The suggested citation for this software is

StataCorp. 2005. *Stata Statistical Software: Release 9.0*. College Station, TX: StataCorp.

Contents

Cross-Referencing the Documentation

When reading this manual, you will find references to other Stata manuals. For example,

      [U]   **26 Overview of Stata estimation commands**
      [XT]   **xtabond**
      [D]   **reshape**

The first is a reference to chapter 26, *Overview of Stata estimation commands* in the *Stata User's Guide*, the second is a reference to the `xtabond` entry in the *Stata Longitudinal/Panel Data Reference Manual*, and the third is a reference to the `reshape` entry in the *Stata Data Management Reference Manual*.

All the manuals in the Stata Documentation have a shorthand notation, such as [U] for the *User's Guide* and [P] for the *Stata Programming Reference Manual*.

The complete list of shorthand notations and manuals is as follows:

[GSW]	*Getting Started with Stata for Windows*
[GSM]	*Getting Started with Stata for Macintosh*
[GSU]	*Getting Started with Stata for Unix*
[U]	*Stata User's Guide*
[R]	*Stata Base Reference Manual*
[D]	*Stata Data Management Reference Manual*
[G]	*Stata Graphics Reference Manual*
[P]	*Stata Programming Reference Manual*
[XT]	*Stata Longitudinal/Panel Data Reference Manual*
[MV]	*Stata Multivariate Statistics Reference Manual*
[SVY]	*Stata Survey Data Reference Manual*
[ST]	*Stata Survival Analysis and Epidemiological Tables Reference Manual*
[TS]	*Stata Time-Series Reference Manual*
[I]	*Stata Quick Reference and Index*
[M]	*Mata Reference Manual*

Detailed information about each of these manuals may be found online at

http://www.stata-press.com/manuals/

About this manual

This manual discusses Stata for Windows®. Stata for Macintosh® users should see *Getting Started with Stata for Macintosh*; Stata for Unix® users should see *Getting Started with Stata for Unix*.

This manual is intended both for people completely new to Stata and for experienced Stata users new to Stata for Windows. Previous Stata users will also find it helpful as a tutorial on some new features in Stata for Windows.

Each numbered chapter opens with a short summary of the contents of that chapter. Previous users will find the summaries a useful reference for Stata's basic commands. We recommend that new users read the summaries and then the more detailed information that follows.

Following the numbered chapters are four appendices with information specific to Stata for Windows.

We provide several types of technical support to registered Stata users. Chapter 6 of this manual describes the resources available to help you learn about Stata's commands and features. One of these resources is the Stata web site (*http://www.stata.com*), where you will find answers to frequently asked questions (FAQs), as well as much other useful information. If you still have questions after looking at the Stata web site and the other resources described in chapter 6, you can contact us as described in [U] **3.9 Technical support**.

1 Installation

Stata for Windows

Description   Stata for Windows is available for Windows 2000 and XP (32-bit and 64-bit x86-64 and Itanium®).

(Versions for Macintosh and for Unix are also available.)

Stata/SE   Professional version of Stata.
Fastest.
Maximum of 32,766 variables; observations limited only by computer memory.
String variables up to 244 characters.
Matrices up to 11,000 x 11,000 in Stata. In Mata, only limited by memory.

Intercooled Stata   Professional version of Stata.
Very fast.
Maximum of 2,047 variables; observations limited only by computer memory.
String variables up to 80 characters.
Matrices up to 800 x 800 in Stata. In Mata, only limited by memory.

Small Stata   Stata for small computers.
Maximum of 99 variables and approximately 1,000 observations.
Slower than Stata/SE or Intercooled Stata.
String variables up to 80 characters.
Matrices up to 40 x 40.

(*Continued on next page*)

Stata for Windows, continued

You received   **Whether you purchased Stata/SE, Intercooled Stata, or Small Stata:**
Single sheet *License and Authorization Key.*
Installation CD.
Registration card.

A Base Stata Documentation Set includes:
Getting Started with Stata for Windows (this book).
Stata User's Guide.
3-volume *Stata Base Reference Manual.*
Stata Data Management Reference Manual.
Stata Graphics Reference Manual.
Stata Quick Reference and Index.

A full Stata Documentation Set includes:
Getting Started with Stata for Windows (this book).
Stata User's Guide.
3-volume *Stata Base Reference Manual.*
Stata Data Management Reference Manual.
Stata Graphics Reference Manual.
Stata Programming Reference Manual.
Stata Multivariate Statistics Reference Manual.
Stata Longitudinal/Panel Data Reference Manual.
Stata Survey Data Reference Manual.
Stata Survival Analysis and Epidemiological Tables Reference Manual.
Stata Time-Series Reference Manual.
Stata Quick Reference and Index.

Additionally, you may have elected to purchase
Mata Reference Manual.

What to install   Look at your *License and Authorization Key.*

If you have a Stata/SE license:
Install Stata/SE.

If you have an Intercooled Stata license:
Install Intercooled Stata.
Do not install Stata/SE; your license will not
let you run it.

If you have a Small Stata license:
Install Small Stata.
Do not install Stata/SE or Intercooled Stata;
your license will not let you run them.

If you have a single-user license, you may install Stata on
both your work computer and your home computer, since it
would not be possible for you to use both at the same time.

Important   Do not lose your paper license. Keep it in a safe place.
You may need it again in the future.

Before you install

Before you begin the installation procedure:

1. Make sure you have the Stata Installation CD.

2. Make sure you have a *License and Authorization Key*.

3. Decide whether you are installing Stata/SE, Intercooled Stata, or Small Stata (see previous page). If you want to install a 64-bit version of Stata, your license must say that it is a 64-bit license and you must perform the installation procedure on a 64-bit version of Windows.

4. Decide where you want to install the Stata software. We recommend `C:\Program Files\Stata9`. If you want to install Stata on a network drive, see [GSW] **D.4 Installing Stata for Windows on a network drive**. If you want to install Stata over an older version, you must uninstall the older version first.

5. Decide where you want to set the working directory. This should be different from the installation directory so that files that you create will not get mixed up with Stata's files. We recommend `C:\data`.

6. If you already have an old version of Stata on your system, decide whether you want to keep it or delete it. If you want to install the new version in the same directory, you must uninstall the old version first. If you want to keep the old version, you must install the new version in a different directory. We do not recommend having more than one version of Stata installed at a time.

Upgrade or update?

If you use Stata 8 or an earlier release and you want to upgrade to Stata 9, this is the chapter for you. If you have already installed Stata 9 and you want to install the latest updates to Stata 9, see chapter 19.

Installation

Be sure that Windows is installed and properly running before you attempt to install Stata for Windows. Have your Stata *License and Authorization Key* with you.

1. Insert the CD in the CD-ROM drive.

2. If you have **Auto-insert Notification** enabled, the installer will start automatically. Otherwise, select **Run** from the Windows **Start** menu, and enter **D:\setup.exe** (assuming that **D:** is the drive letter for your CD-ROM) to start the installer.

3. You will see an opening screen of information. After reading this screen carefully, click **Next**.

4. The installer will display options for personalizing your installation and making Stata accessible to all users that share your computer. Make any necessary changes, and click **Next**.

5. At the **Select Components** dialog, you can choose which flavor of Stata to install.

   a. Look at your *License and Authorization Key*. If you have a Stata/SE license, choose **Stata/SE**.

   b. If you have an Intercooled Stata license, choose **Intercooled Stata**. Stata/SE will not run under an Intercooled Stata license.

   c. If you have a Small Stata license, choose **Small Stata**. Stata/SE and Intercooled Stata will not run under a Small Stata license.

   d. If you ran the installer on a 64-bit version of Windows and your license allows it, you can choose to install a 64-bit version of Stata/SE or Intercooled Stata. Your license must say that it is a 64-bit license if you want to install a 64-bit version of Stata.

6. The installer will ask you where you want to install Stata.

   a. We recommend that you choose the default directory—C:\Program Files\Stata9.

   b. If you want to install Stata on a network drive, see [GSW] **D.4 Installing Stata for Windows on a network drive** before continuing.

   c. If an older version of Stata is already installed in the directory that you have chosen, you must uninstall it first. This prevents out-of-date Stata programs from being mixed in with new ones. If you do not want to uninstall the older version, choose a different directory in which to install Stata.

   d. When you have chosen an installation directory, click **Next**.

7. The installer will then ask you where you want to set the default working directory.

   a. The default working directory is where your datasets, graphs, and other Stata-related files will be stored.

   b. We recommend that you choose the default directory—C:\data.

   c. When you have chosen a default working directory, click **Next** to begin the installation.

8. When the installation is complete, click **Finish** to exit the installer.

Initialize the license

The first time that you start any version of Stata, it will prompt you for the information on your *License and Authorization Key*. You must enter something for all fields in the dialog before you can continue. The code and authorization are not case-sensitive. If you make a mistake typing the codes, you will be prompted to try again.

If you get the message "The serial number, code, and authorization are inconsistent!", try the initialization again. Be careful when typing your code and authorization key. Note that anything that looks like "o" is the letter *oh*, anything like "0" is a zero, anything like "1" is the number one, and anything like "L" is the letter *el*.

Update Stata if necessary

StataCorp releases updates to Stata often. These updates may include new features and bug fixes that can be automatically downloaded and installed by Stata from the Internet. There may be updates to Stata more recent than the version of Stata on your CD.

The first time Stata is launched, a dialog will open asking if you would like to check for updates now. Click **OK** to do so. If an update is available, follow the instructions. If you have trouble connecting to the Internet from Stata, visit *http://www.stata.com/support/faqs/web/* for help.

It's also a good idea to periodically check for updates to Stata. Stata can automatically check for updates for you. See chapter 19 for more information about updating.

Verify the installation

You should verify that the installation was performed correctly. Stata includes a command called `verinst`. Start Stata, type `verinst` in the Command window, and press *Enter*.

You should see output in the Results window informing you that Stata is correctly installed.

```
. verinst
You are running Stata/SE 9.0 for Windows.

Stata is correctly installed.
You can type exit to exit Stata.
```

If you see the error message below complaining that the command was not found,

```
. verinst
unrecognized command
r(199);
```

then Stata is not installed properly; see [GSW] **B.2 verinst problems**.

If Stata informed you that it was correctly installed, you can do one of the following to exit:

1. Select **File > Exit**.

2. Type `exit` in the Command window, and press *Enter*. In the future, you may need to type `exit, clear` if your data has changed since you last saved it.

2 Starting and exiting Stata

Starting Stata

To start Stata:

1. Select it from the Windows **Start** menu.

2. Or double-click on a Stata data file. Stata data files are data files created by Stata and have the extension .dta. When you double-click on a Stata data file, Stata is launched and the data file is loaded into Stata; see chapter 7.

When Stata has launched, you will see

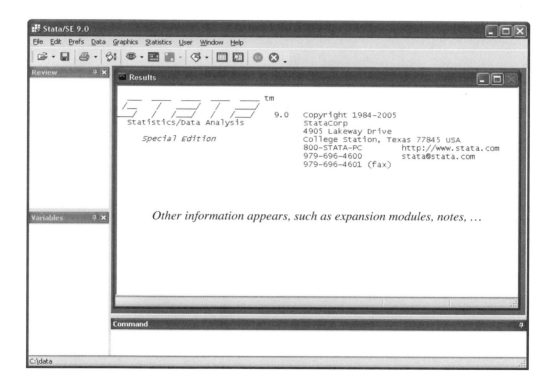

Note

There are other ways to start Stata. You may find one of these ways more convenient. Please read [GSW] **A. More on starting and exiting Stata** at the end of this manual for more information.

Exiting Stata

To exit Stata, enter `exit` from the command line, or select **File > Exit** from Stata's menus.

In the future, you might see the following when you type `exit`:

```
. exit
no; data in memory would be lost
r(4);

. _
```

As you will learn, you will see this when you have changed the dataset in memory and have not saved it. Either you can save the dataset by typing `save` *filename* and then typing `exit`, or you can force Stata to exit without saving the dataset by typing `exit, clear`.

```
. exit, clear
```

If you experienced any problems when trying to start Stata, see [GSW] **B. Troubleshooting starting and exiting Stata**.

3 A sample session

Introducing Stata

In this chapter, you will learn:

Most of Stata's commands share a common syntax, which is

$[\text{prefix\_cmd}:]$ command $[\text{varlist}]$ $[\text{if}]$ $[\text{in}]$ $[, \text{options}]$

where items enclosed in square brackets are optional.

Example:
```
list
list mpg weight
list if mpg>20
list mpg weight if mpg>20
list in 1/10
list mpg weight in 1/10
list mpg weight if mpg>20 in 1/10
```

Most Stata commands also allow some prefix commands; see [U] **11.1.10 Prefix commands**. For example, the prefix command by repeats the following command for each set of values of the by *varlist*:
```
summarize mpg
by rep78: summarize mpg
```
The first command reports the mean, etc., of mpg.
The second command reports the mean, etc., of mpg for each value of rep78.

Determine which are which by selecting **Help > Stata Command...**,
entering *commandname*, and examining the syntax diagram in the help file.

See **Help** for summarize and discover that by is allowed.
See **Help** for label and discover that by is not allowed.

A single comma sets off a command's options from the rest of the command.
```
list mpg weight, divider separator(3)
```
is the correct way to specify options divider and separator().
```
list mpg weight, divider, separator(3)
```
with commas between the options is incorrect.

Warning:

The purpose of this sample session is to show you a little of Stata in action.

Please do not assume that because we do not demonstrate a feature,
Stata cannot do it.
Stata has lots of other commands.

Please do not assume that because we demonstrate a command,
we demonstrate all aspects of that command.
Stata has lots of options.

See the Reference manuals for a complete description of Stata's features.
When in doubt, consult the indices located at the back of the manuals.

Sample session

For this chapter, you should be sitting at your computer. We are going to introduce Stata by leading you through a sample session. We will use the `auto.dta` file, which was shipped with Stata. If you wish to follow along, type the commands that follow a dot (`.`) throughout the session.

```
. sysuse auto
(1978 Automobile Data)
```

Alternatively, you may load the `auto.dta` file directly from the Stata web site; you will learn more about this in chapter 19.

The data that we loaded contain

```
. describe
Contains data from C:\Program Files\Stata9\ado\base\a\auto.dta
  obs:            74                          1978 Automobile Data
 vars:            12                          13 Apr 2005 17:45
 size:         3,478 (99.5% of memory free)   (_dta has notes)

              storage  display     value
variable name   type   format      label      variable label

make           str18   %-18s                   Make and Model
price          int     %8.0gc                  Price
mpg            int     %8.0g                   Mileage (mpg)
rep78          int     %8.0g                   Repair Record 1978
headroom       float   %6.1f                   Headroom (in.)
trunk          int     %8.0g                   Trunk space (cu. ft.)
weight         int     %8.0gc                  Weight (lbs.)
length         int     %8.0g                   Length (in.)
turn           int     %8.0g                   Turn Circle (ft.)
displacement   int     %8.0g                   Displacement (cu. in.)
gear_ratio     float   %6.2f                   Gear Ratio
foreign        byte    %8.0g       origin      Car type

Sorted by: foreign
```

Note that you can also use Stata's dialogs to perform these operations by selecting the menus and dialogs corresponding to the commands. To perform the above two commands using Stata's menus and dialogs,

1. Select **File > Example Datasets...**.
2. Click on **Example datasets installed with Stata**.
3. Click on **use** for `auto.dta`.
4. Select **Data > Describe data > Describe variables in memory**.
5. Leave the **Variables** field empty to indicate that you want all variables.
6. Click **OK**.

Listing can be informative

Here are some of our data:

```
. list make mpg in 1/10
```

	make	mpg
1.	AMC Concord	22
2.	AMC Pacer	17
3.	AMC Spirit	22
4.	Buick Century	20
5.	Buick Electra	15
6.	Buick LeSabre	18
7.	Buick Opel	26
8.	Buick Regal	20
9.	Buick Riviera	16
10.	Buick Skylark	19

To do this with a dialog,

1. Select **Data > Describe data > List data**.
2. Click the button to the right of the **Variables** field, and select make to enter it into the **Variables** field.
3. Click the button to the right of the **Variables** field, and select mpg to append it to the **Variables** field.
4. Click the **by/if/in** tab in the dialog.
5. Check the **Use a range of observations** checkbox.
6. Type 1 in the **From:** field and 10 in the **to:** field.
7. Click **Submit**.

Note

The **Variables** field is actually a combobox control that accepts variable names. Clicking the button to the right of the **Variables** edit field displays a list of the variables from the current dataset. Depending on the control, selecting a variable from the list will either enter the variable name into the edit field, replace the contents of the edit field with the variable name, or append the variable name to the contents of the edit field.

You can click the arrows to enter the values for the **From:** and **to:** fields instead of typing them in. Clicking **Submit** instead of **OK** submits the command to Stata without closing the dialog.

Aside: Typing browse or edit instead of list brings up the requested data in a spreadsheet-style editor for reading or entering data, respectively.

Listing can be informative, continued

Question: Which cars yield the lowest gas mileage?

```
. sort mpg
. l make mpg in 1/5

       +----------------------+
       | make            mpg  |
       |----------------------|
  1.   | Linc. Mark V     12  |
  2.   | Linc. Continental 12 |
  3.   | Merc. Cougar     14  |
  4.   | Merc. XR-7       14  |
  5.   | Linc. Versailles 14  |
       +----------------------+
```

Note that we abbreviated `list` as `l`. Many Stata commands allow abbreviations. If you look at the manual entry for `list`, you will see the minimum abbreviation for it designated by `list`.

Which 5 cars yield the highest gas mileage?

```
. l make mpg in -5/-1

        +---------------------+
        | make           mpg  |
        |---------------------|
  70.   | Toyota Corolla  31  |
  71.   | Plym. Champ     34  |
  72.   | Subaru          35  |
  73.   | Datsun 210      35  |
  74.   | VW Diesel       41  |
        +---------------------+
```

Note

`-5` means the fifth from the last observation, and `-1` means the last observation. See [U] **11.1.4 in range** for more information.

Descriptive statistics

Question: What is the average price of a car in these data?

```
. summarize price
    Variable │      Obs        Mean    Std. Dev.       Min        Max
─────────────┼───────────────────────────────────────────────────────
       price │       74    6165.257    2949.496       3291      15906
```

Aside: summarize works like list—without arguments it provides a summary of all the data:

```
. summarize
    Variable │      Obs        Mean    Std. Dev.       Min        Max
─────────────┼───────────────────────────────────────────────────────
        make │        0
       price │       74    6165.257    2949.496       3291      15906
         mpg │       74     21.2973    5.785503         12         41
       rep78 │       69    3.405797    .9899323          1          5
    headroom │       74    2.993243    .8459948        1.5          5
─────────────┼───────────────────────────────────────────────────────
       trunk │       74    13.75676    4.277404          5         23
      weight │       74    3019.459    777.1936       1760       4840
      length │       74    187.9324    22.26634        142        233
        turn │       74    39.64865    4.399354         31         51
displacement │       74    197.2973    91.83722         79        425
─────────────┼───────────────────────────────────────────────────────
  gear_ratio │       74    3.014865    .4562871       2.19       3.89
     foreign │       74    .2972973    .4601885          0          1
```

Note

make has 0 observations because it is a string—calculating a mean is undefined but is not an error. rep78 has only 69 observations because for five cars, it is missing.

Descriptive statistics, continued

Question: What is the average price of cars that are below and above the mean MPG?

```
. summarize price if mpg<21.3
    Variable |      Obs        Mean    Std. Dev.       Min        Max

       price |       43     7091.86    3425.019       3291      15906
. summarize price if mpg>=21.3
    Variable |      Obs        Mean    Std. Dev.       Min        Max

       price |       31    4879.968    1344.659       3299       9735
```

Aside: `if` can be suffixed to almost all commands. This is one of Stata's more useful features. See [U] **11.1.3 if exp** for more information.

Question: What is the median MPG?

```
. summarize mpg, detail
                        Mileage (mpg)

         Percentiles     Smallest
  1%          12             12
  5%          14             12
 10%          14             14       Obs                 74
 25%          18             14       Sum of Wgt.         74

 50%          20                      Mean           21.2973
                          Largest     Std. Dev.     5.785503
 75%          25             34
 90%          29             35       Variance      33.47205
 95%          34             35       Skewness      .9487176
 99%          41             41       Kurtosis      3.975005
```

Answer: 20.

Descriptive statistics, continued

Aside: The ', detail' at the end of the summarize command is called an option. Most Stata commands share a common syntax:

$$[prefix\_cmd:]\ command\ [varlist]\ \big[\,if\,\big]\ \big[\,in\,\big]\ [\,,\ options\,]$$

Square brackets mean something is optional. Thus,

command by itself is valid:	summarize
command followed by a *varlist* (variable list) is valid:	summarize mpg
	summarize mpg weight
command with if (with or without a *varlist*) is valid:	summarize if mpg>20
	summarize mpg weight if mpg>20

and so on.

If *varlist* is not specified, all the variables are used.

if and in restrict the data on which the command is run.

options modify what the command does.

Most commands also accept prefix commands that modify their behavior; see [U] **11.1.10 Prefix commands** for details. One of the more common prefix commands is by.

Each command's syntax is found in the Reference manuals. You can learn about summarize in [R] **summarize**, or select **Help > Stata Command...** and enter summarize.

Descriptive statistics, continued

Our dataset contains variable `foreign`, which is 0 if the car was manufactured in the United States or Canada and 1 otherwise.

Problem: Obtain summary statistics for price and MPG for each value of foreign.

There are two solutions to this problem:

1. We could type in the commands:

   ```
   summarize price mpg if foreign==0
   summarize price mpg if foreign==1
   ```

2. Or, we could do the following:

```
. sort foreign
. by foreign: summarize price mpg

-> foreign = Domestic
    Variable |      Obs        Mean   Std. Dev.       Min       Max

       price |       52    6072.423    3097.104      3291     15906
         mpg |       52    19.82692    4.743297        12        34

-> foreign = Foreign
    Variable |      Obs        Mean   Std. Dev.       Min       Max

       price |       22    6384.682    2621.915      3748     12990
         mpg |       22    24.77273    6.611187        14        41
```

Descriptive statistics, continued

Explanation: Generally, Stata's full command syntax is

$$[prefix\_cmd:] \; command \; [varlist] \; \big[\, if \, \big] \; \big[\, in \, \big] \; \big[\, , \; options \, \big]$$

When the *prefix_cmd* command by is placed in front of a command, the command is repeated for each set of values of the variables specified after by.

Most commands allow by, but some do not. Online help always begins with a command's syntax diagram. So, you can quickly tell whether by is allowed by clicking on **Help**, selecting **Stata Command...**, and entering *commandname*. If by is allowed, you will see

by may be used with *commandname*; see help by

immediately following the options list.

More explanation: Actually, Stata's command syntax includes even more than we have shown you. See [U] **11 Language syntax**.

Note on by: To use by, you need to sort the data by the by-variables. That is why we typed sort foreign before typing by foreign: .... Additional features of the by prefix, such as automatic sorting, can be found in [D] **by**.

Minor note: How is it that foreign takes on values 0 and 1, and yet in the output, its values appear to be "Foreign" and "Domestic"? foreign has a value label, associating 0 with Domestic and 1 with Foreign; see chapter 10.

More on by

Problem: It appears that the average MPG of domestic and foreign cars differs. We want to test the hypothesis that the means are equal.

```
. ttest mpg, by(foreign)

Two-sample t test with equal variances

    Group |     Obs        Mean    Std. Err.    Std. Dev.   [95% Conf. Interval]
----------+--------------------------------------------------------------------
 Domestic |      52    19.82692     .657777     4.743297    18.50638    21.14747
  Foreign |      22    24.77273     1.40951     6.611187    21.84149    27.70396
----------+--------------------------------------------------------------------
 combined |      74     21.2973    .6725511     5.785503     19.9569    22.63769
----------+--------------------------------------------------------------------
     diff |            -4.945804    1.362162                -7.661225   -2.230384
--------------------------------------------------------------------------------
    diff = mean(Domestic) - mean(Foreign)                          t =  -3.6308
Ho: diff = 0                                      degrees of freedom =        72

    Ha: diff < 0                 Ha: diff != 0                   Ha: diff > 0
 Pr(T < t) = 0.0003       Pr(|T| > |t|) = 0.0005            Pr(T > t) = 0.9997
```

Syntax explanation: The by in this command is not the by prefix described on the previous page; it is an option that just happens to be named by().

For the ttest command, the by() option informs the command what groups to test for equality. ttest mpg, by(foreign) tests that the means of mpg in the groups defined by foreign are equal.

Another aside: We showed only one of the three syntaxes for ttest. See the help file for ttest or [R] **ttest** for a complete description of ttest.

Analysis note: We have established that in 1978 domestic cars had poorer gas mileage than foreign cars.

Descriptive statistics, making tables

Problem: Obtain counts of the number of domestic and foreign cars.

```
. tabulate foreign
     Car type |      Freq.     Percent        Cum.
--------------+-----------------------------------
     Domestic |         52       70.27       70.27
      Foreign |         22       29.73      100.00
--------------+-----------------------------------
        Total |         74      100.00
```

Problem: The dataset contains variable rep78 recording each car's frequency-of-repair record (1 = poor, ..., 5 = excellent). Obtain frequency counts.

```
. tabulate rep78
       Repair |
  Record 1978 |      Freq.     Percent        Cum.
--------------+-----------------------------------
            1 |          2        2.90        2.90
            2 |          8       11.59       14.49
            3 |         30       43.48       57.97
            4 |         18       26.09       84.06
            5 |         11       15.94      100.00
--------------+-----------------------------------
        Total |         69      100.00
```

Problem: We have 74 cars; only 69 have frequency-of-repair records recorded. List the cars for which data is missing.

```
. list make if missing(rep78)

     +---------------+
     |         make  |
     |---------------|
  3. |    AMC Spirit |
  7. |    Buick Opel |
 45. | Plym. Sapporo |
 51. | Pont. Phoenix |
 64. |   Peugeot 604 |
     +---------------+
```

Note list make if rep78>=. is equivalent to list make if missing(rep78); Stata stores missing values as the largest possible numeric values. See [U] **12.2.1 Missing values** for full details.

Descriptive statistics, making tables, continued

Problem: Compare frequency-of-repair records for domestic and foreign cars (i.e., make a two-way table).

```
. tabulate rep78 foreign
   Repair
   Record           Car type
     1978     Domestic     Foreign  |    Total

        1            2           0  |        2
        2            8           0  |        8
        3           27           3  |       30
        4            9           9  |       18
        5            2           9  |       11

    Total           48          21  |       69
```

Problem: Domestic cars appear to have poorer frequency-of-repair records. Is the difference statistically significant? Obtain a χ^2 (even though there are not at least 5 cars expected in each cell):

```
. tabulate rep78 foreign, chi2
   Repair
   Record           Car type
     1978     Domestic     Foreign  |    Total

        1            2           0  |        2
        2            8           0  |        8
        3           27           3  |       30
        4            9           9  |       18
        5            2           9  |       11

    Total           48          21  |       69
          Pearson chi2(4) =   27.2640    Pr = 0.000
```

Analysis note: We find that frequency-of-repair records differ between domestic and foreign cars. In 1978, domestic cars appear to be poorer in this regard.

Descriptive statistics, making tables, continued

Aside: `tabulate` provides options to display any or all of the row percentages, column percentages, and cell percentages along with, or instead of, the frequencies.

The option `nofreq` suppresses the frequencies; option `row` adds row percentages; option `col` adds column percentages; and option `cell` adds cell percentages.

```
. tabulate rep78 foreign, chi2 row col

  ┌─────────────────┐
  │ Key             │
  ├─────────────────┤
  │      frequency  │
  │  row percentage │
  │ column percentage │
  └─────────────────┘

  Repair │
  Record │        Car type
    1978 │   Domestic     Foreign │     Total
  ───────┼──────────────────────┼──────────
       1 │         2          0  │         2
         │    100.00       0.00  │    100.00
         │      4.17       0.00  │      2.90
  ───────┼──────────────────────┼──────────
       2 │         8          0  │         8
         │    100.00       0.00  │    100.00
         │     16.67       0.00  │     11.59
  ───────┼──────────────────────┼──────────
       3 │        27          3  │        30
         │     90.00      10.00  │    100.00
         │     56.25      14.29  │     43.48
  ───────┼──────────────────────┼──────────
       4 │         9          9  │        18
         │     50.00      50.00  │    100.00
         │     18.75      42.86  │     26.09
  ───────┼──────────────────────┼──────────
       5 │         2          9  │        11
         │     18.18      81.82  │    100.00
         │      4.17      42.86  │     15.94
  ───────┼──────────────────────┼──────────
   Total │        48         21  │        69
         │     69.57      30.43  │    100.00
         │    100.00     100.00  │    100.00

        Pearson chi2(4) =  27.2640   Pr = 0.000
```

Note

We type `tabulate ..., chi2 row col`, not `tabulate ..., chi2, row, col`. One comma sets off the options from the command; commas do not separate options from each other. The order of the options is irrelevant.

Descriptive statistics, correlation matrices

Question: What is the correlation between MPG and weight of car?

```
. correlate mpg weight
(obs=74)
                 mpg    weight

       mpg      1.0000
    weight     -0.8072    1.0000
```

Problem: Compare the correlation for domestic and foreign cars.

```
. correlate mpg weight if foreign==0
(obs=52)
                 mpg    weight

       mpg      1.0000
    weight     -0.8759    1.0000

. correlate mpg weight if foreign==1
(obs=22)
                 mpg    weight

       mpg      1.0000
    weight     -0.6829    1.0000
```

Note

We could have obtained this by typing by foreign: correlate mpg weight instead.

Aside: We can produce correlation matrices containing as many variables as we wish.

```
. correlate mpg weight price length displacement
(obs=74)
                    mpg    weight     price    length  displa~t

          mpg      1.0000
       weight     -0.8072    1.0000
        price     -0.4686    0.5386    1.0000
       length     -0.7958    0.9460    0.4318    1.0000
 displacement     -0.7056    0.8949    0.4949    0.8351    1.0000
```

Graphing data

Problem: We know that the average MPG of domestic and foreign cars differs. We have learned that domestic and foreign cars differ in other ways as well, such as in frequency-of-repair record. We found a negative correlation of MPG and weight—as we would expect—but the correlation appears stronger for domestic cars. Examine, with an eye toward modeling, the relationship between MPG and weight. Begin with a graph.

```
.  scatter mpg weight
```

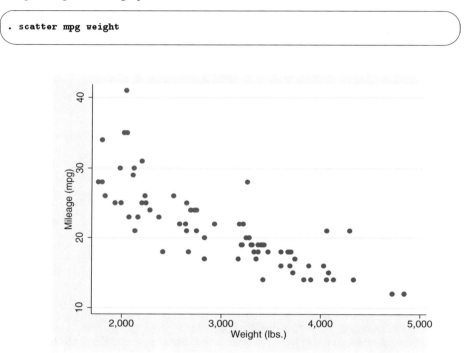

Comment: `scatter` is explained in the *Graphics Reference Manual*, but typing `scatter` y x draws a graph of y against x. The relationship, we note, appears to be nonlinear.

Graphing data, continued

Next we draw separate graphs for foreign and domestic cars.

```
. scatter mpg weight, by(foreign, total)
```

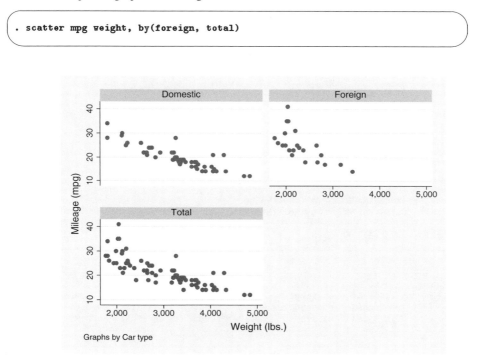

Syntax note: by() is on the right of the command; therefore, scatter did whatever it is that it does with the grouping information. What scatter did is draw separate graphs for domestic and foreign cars in a single image. We have only two groups, but scatter will allow any number—the individual graphs just get smaller. The total option within by() added an overall graph to the image.

To create the graph with a dialog,

1. Select **Graphics > Twoway graph (scatterplot, line, etc.)**.
2. Choose **Scatter** for the plot type.
3. Enter weight in the **X variable** field and mpg in the **Y variable** field.
4. Click the **By** tab.
5. Enter foreign in the **Variables** field.
6. Check the **Graph total** checkbox.
7. Click **OK**.

Analysis note: The relationship is not only nonlinear; the domestic-car relationship appears to differ from that of foreign cars.

Model fitting: linear regression

Restatement of problem: We are to model the relationship between MPG and weight.

Plan of attack: Based on the graphs, we judge the relationship to be nonlinear and will model MPG as a quadratic in weight. Also based on the graphs, we judge the relationship to be different for domestic and foreign cars. We will include an indicator (dummy) variable for foreign and evaluate afterwards whether this adequately describes the difference. Thus, we will fit the model

$$\text{mpg} = \beta_0 + \beta_1\,\text{weight} + \beta_2\,\text{weight}^2 + \beta_3\,\text{foreign} + \epsilon$$

foreign is already a 0/1 variable, so we only need to create the weight-squared variable:

```
. generate wtsq = weight^2
. regress mpg weight wtsq foreign
```

Source	SS	df	MS		
Model	1689.15372	3	563.05124		
Residual	754.30574	70	10.7757963		
Total	2443.45946	73	33.4720474		

Number of obs = 74
F(3, 70) = 52.25
Prob > F = 0.0000
R-squared = 0.6913
Adj R-squared = 0.6781
Root MSE = 3.2827

mpg	Coef.	Std. Err.	t	P>|t|	[95% Conf. Interval]	
weight	-.0165729	.0039692	-4.18	0.000	-.0244892	-.0086567
wtsq	1.59e-06	6.25e-07	2.55	0.013	3.45e-07	2.84e-06
foreign	-2.2035	1.059246	-2.08	0.041	-4.3161	-.0909002
_cons	56.53884	6.197383	9.12	0.000	44.17855	68.89913

Aside: Stata can fit many kinds of models, including logistic regression, Cox proportional hazards, etc. Type help estimation commands for a complete list.

Continuation of attack: We obtain the predicted values:

```
. predict mpghat
(option xb assumed; fitted values)
```

Comment: Be sure to read [U] **20 Estimation and postestimation commands**. There are a number of features available to you after estimation—one is calculating predicted values. predict just created a new variable called mpghat equal to

$$-.0165729\,\text{weight} + 1.59 \times 10^{-6}\,\text{wtsq} - 2.2035\,\text{foreign} + 56.53884$$

Model fitting: linear regression, continued

We can now graph the data and the predicted curve.

Continuation of attack: We just created `mpghat` with `predict`. We could graph the fit and data, but we want to evaluate the fit on the foreign and domestic data separately to determine if our shift parameter is adequate. We can draw both graphs using one command with the `by()` option:

```
. sort weight
. scatter mpg weight || line mpghat weight, by(foreign)
```

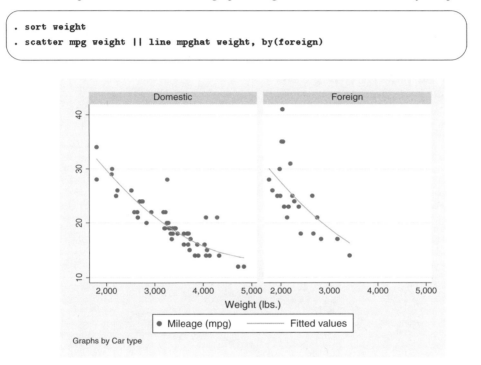

Graphs by Car type

Our `scatter` command looks much like the one we used before our regression, but we have added `|| line mpghat weight` and left off the `total` suboption of `by()`. The syntax of the `scatter` command is an extended form of what we have seen; it allows for multiple plots to be overlaid by specifying each plot after a `||` *plot command*. Here we have overlaid a line plot of our predictions, `mpghat`, atop our original scatters.

To create the graph with a dialog,

1. Select **Graphics > Overlaid twoway graphs**.
2. From the **Plot 1** tab, choose **Scatter** for the plot type.
3. Enter `weight` in the **X variable** field and `mpg` in the **Y variable** field.
4. Click the **Plot 2** tab, and choose **Line** for the plot type.
5. Enter `weight` in the **X variable** field and `mpghat` in the **Y variable** field.
6. Click the **By** tab.
7. Enter `foreign` in the **Variables** field.
8. Click **OK**.

Model fitting: linear regression, continued

Problem: We show our results to an engineer. "No," he says. "It should take twice as much energy to move 2,000 pounds 1 mile compared with moving 1,000 pounds, and therefore twice as much gasoline. Miles per gallon is not a quadratic in weight, gallons per mile is a linear function of weight."

We go back to the computer:

```
. gen gpm = 1/mpg
. label var gpm "Gallons per mile"
. sort foreign
. scatter gpm weight, by(foreign, total)
```

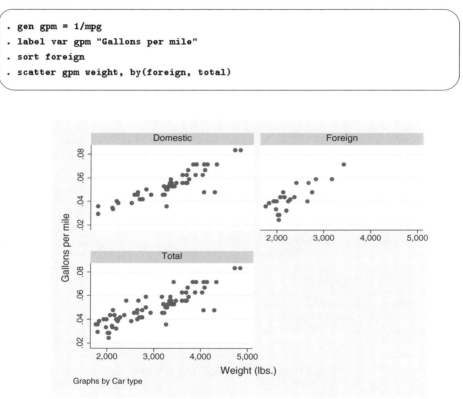

Model fitting: linear regression, continued

Satisfied that the engineer is indeed correct, we rerun the regression:

```
. regress gpm weight foreign
      Source |       SS       df       MS              Number of obs =      74
-------------+------------------------------           F(  2,    71) =  113.97
       Model |  .009117618      2  .004558809           Prob > F      =  0.0000
    Residual |   .00284001     71      .00004           R-squared     =  0.7625
-------------+------------------------------           Adj R-squared =  0.7558
       Total |  .011957628     73  .000163803           Root MSE      =  .00632

         gpm |      Coef.   Std. Err.       t    P>|t|     [95% Conf. Interval]
-------------+----------------------------------------------------------------
      weight |   .0000163   1.18e-06    13.74   0.000     .0000139    .0000186
     foreign |   .0062205   .0019974     3.11   0.003     .0022379    .0102032
       _cons |  -.0007348   .0040199    -0.18   0.855    -.0087504    .0072807
```

We find foreign cars in 1978 to be less efficient. Foreign cars may have yielded better gas mileage than domestic cars in 1978, but this was only because they were so light.

Notes

4 The Stata user interface

In this chapter, you will learn:

The Stata windows:
- Results are displayed in the Results window
- Commands are entered into the command line from the Command window
- Previously entered commands are displayed in the Review window
- The variables for the current dataset are displayed in the Variables window
- The Variables window can be used to enter variable names into the Command window
- A window can be opened by selecting it from the **Window** menu
- Pressing *Ctrl-Tab* cycles through all open windows in Stata

The Stata toolbar:
- Provides access to Stata's more commonly used features
- Holding the mouse pointer over a button displays a description of the button
- A window can be opened by clicking its button on the toolbar

Right-clicking on a window displays options for that window

Menus and dialogs provide access to commands in Stata

The working directory is displayed on the status bar

The windows

The four main windows are the Review, Variables, Results, and Command window, and are initially contained within the main Stata window. The menus and toolbar provide access to Stata's dialogs and utilities, such as the Viewer, Do-file Editor, and Data Editor.

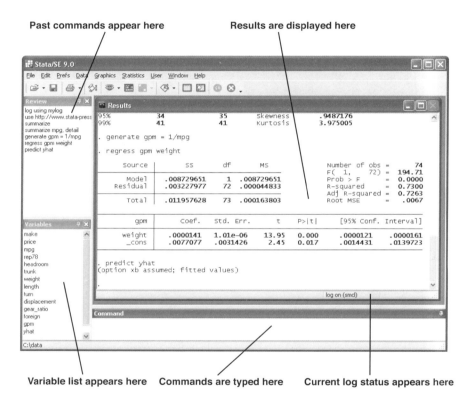

Past commands appear here   Results are displayed here

Variable list appears here   Commands are typed here   Current log status appears here

To open a window such as the Viewer, select the window from the **Window** menu or the toolbar. If a window is open but behind other windows, you can bring the window to the front by selecting it from the **Window** menu or the toolbar. Or you can press *Ctrl-Tab* to cycle through all open windows in Stata.

Many of Stata's windows have functionality that can be accessed by clicking the right mouse button (right-clicking) on the window. Doing so displays a contextual menu that, depending on the window, allows you to copy text, set the preferences for the window, or print the contents of the window. When copying text or printing, we recommend that you always right-click on the window rather than use the main menu or toolbar so that you can be sure of where and what you're copying or printing.

The toolbar

The toolbar contains buttons that provide quick access to Stata's more commonly used features. If you ever forget what a button does, hold the mouse pointer over a button for a moment and a window will appear with a description of that button. Some buttons contain both an icon and an arrow. Clicking on the icon performs a default action, while clicking on the arrow displays a menu.

Open

> Open a Stata dataset. Click on the icon to open a dataset with the **Open** dialog. Click on the arrow to select a dataset from a menu of recently opened datasets.

Save

> Save the Stata dataset currently in memory to disk.

Print

> Print Results, Viewer, or Graph. Click on the icon to print the contents of the Results window. Click on the arrow to select a window to print.

Log Begin/Close/Suspend

> Begin a new log, append to an existing log, or close or suspend the current log. See chapter 16 for an explanation of log files.

Viewer

> Open the Viewer, or bring a Viewer to the front of all other windows. Click on the icon to open a new viewer. Click on the arrow to select a Viewer to bring to the front. See chapter 5 for more information.

Bring Results to Front

> Bring the Results window to the front of all other windows.

The toolbar, continued

Graph

Bring a graph window to the front of all other windows. Click on the icon to bring the topmost graph window to the front. Click on the arrow to select a graph window to bring to the front. See chapter 15 for more information.

Do-file Editor

Open the Do-file Editor, or bring the Do-file Editor to the front of all other windows. Click on the icon to open the Do-file Editor. Click on the arrow to select a Do-file Editor to bring to the front. See chapter 14 for more information.

Data Editor

Open the Data Editor, or bring the Data Editor to the front of the other Stata windows. See chapter 8 for more information.

Data Browser

Open the Data Browser, or bring the Data Browser to the front of the other Stata windows. See chapter 8 for more information.

Clear —more— Condition

Tell Stata to continue when it has paused in the middle of long output. See chapter 11 for more information.

Break

Stop the current task in Stata. See chapter 11 for more information.

The Command window

Commands are submitted to Stata from the command line. The command line is contained within the Command window and supports basic text editing, copying and pasting, a command history, function-key mapping, and variable-name completion. From the command line, pressing

Page Up	Steps backwards through the command history
Page Down	Steps forward through the command history
Tab	Auto-completes a partially typed variable name

See [U] **10 Keyboard use** for more information about keyboard shortcuts for the command line.

The command history allows you to recall a previously submitted command, edit it if you wish, and then resubmit it. Commands submitted by Stata's dialogs are also included in the command history, so you can recall and submit a command without having to open the dialog again.

The Review window

The Review window shows the history of commands that have been entered.

To enter a command from the Review window
- Click once on a past command to copy it to the Command window.
- Double-click on a past command to copy it to the Command window and execute it.

Right-clicking on the Review window displays a menu from which you can select
- **Save Review Contents...** to save the contents of the Review window to a text file.
- **Copy Review Contents to Clipboard** to copy the contents of the Review window to the Clipboard.
- **Font...** to change the font for the window.

The Variables window

The Variables window shows the list of variables and their labels for the dataset in memory.

To enter a variable from the Variables window
- Click once on a variable to copy it to the Command window.
- Double-click on a variable and the variable will be copied twice.

Right-clicking on the Variables window displays a menu from which you can select
- **Define Notes for Variable '*varname*'...** to open the **Notes** dialog for variable *varname*.
- **Font...** to change the font for the window.

Menus and dialogs

Stata is, at its heart, a command-driven application. That is, you issue commands to Stata to tell it to do things, such as load a dataset or perform a regression. As you become familiar with Stata's commands, you will find that they give Stata great power and flexibility.

You can type commands by hand (advanced users actually prefer this). You can also access most of Stata's commands by selecting items from Stata's menus to open dialogs that build Stata commands.

Stata's **Data**, **Graphics**, and **Statistics** menus provide point-and-click access to almost every command in Stata. As you will learn, Stata is fully programmable, and Stata programmers can even create their own dialogs and menus. The **User** menu provides a place for programmers to add their own menu items. Initially, it contains only some empty submenus.

If you wish to perform a Poisson regression, for example, you could type Stata's `poisson` command, or you could select **Statistics > Count outcomes > Poisson regression**, which would display this dialog:

This dialog provides access to all of the functionality of Stata's `poisson` command. The `poisson` command has many options that can be accessed by clicking the multiple tabs across the top of the dialog. The first time you use the dialog for a command, it is a good idea to look at the contents of each tab so that you will know all the dialog's capabilities.

The dialogs for many commands have the **by/if/in** and **Weights** tabs. These provide access to Stata's functions for controlling the estimation sample and dealing with weighted data. See [U] **11 Language syntax** for more information on these features of Stata's language.

The dialogs for most estimation commands have the **Max Options** tab for setting the maximization options (see [R] **maximize**). For example, you can specify the maximum number of iterations for the optimizer.

Menus and dialogs, continued

Most dialogs in Stata provide the same six buttons you see at the bottom of the Poisson dialog on the previous page: **OK**, **Cancel**, **Submit**, **?**, **R**, and **Copy**. **OK**, **Cancel**, and **Submit** do what you would expect.

OK dismisses the dialog and issues a Stata command based on how you have filled out the fields in the dialog. **Cancel** dismisses the dialog without doing anything. **Submit** issues a command just like **OK** but leaves the dialog on the screen so that you can make changes and issue another command.

The command issued by a dialog is submitted just as if you typed it by hand. You can see it in the Results window and the Review window after it executes; this will help you learn Stata's command syntax.

The button in the lower-left corner of the dialog with a question mark on it provides access to Stata's help system. Clicking this button will typically take you to the help file for the Stata command associated with the dialog. In this case, clicking on it would take you to the `poisson` help file. The help file will have have tabs above groups of options to show which dialog tab contains those options.

The button with an **R** on it next to the help button is the "reset" button. Each time you open a dialog, it will remember how you last filled it out. If you wish to reset its fields to their default values at any time, simply click this button.

The button with a clipboard icon on it next to the reset button is the "copy" button. The copy button behaves much like the **Submit** button, but rather than issue a command, it copies the command to the Clipboard. The command can then be pasted elsewhere (such as the Do-file Editor for example).

In addition to being able to access the dialogs for Stata commands through Stata's menus, you can also invoke them using two other methods. You may know the name of a Stata command for which you want to see a dialog, but you may not remember how to navigate to that command in the menu system. Simply type db *commandname* to launch the dialog for *commandname*:

    . db poisson

You will also find access to the dialog for a command in that command's help file; see chapter 6 for more details.

As you read this manual, we will present examples of Stata commands. You may type those examples as presented, but you should also experiment with submitting those commands by using their dialogs. Use the db command described above to quickly launch the dialog for any command that you see in this manual.

The working directory

If you look at the screenshot on page 32, you will notice a status bar at the bottom of the main window that contains C:\data. This indicates that C:\data is the current working directory.

The working directory is where graphs and datasets will be saved unless you specify another directory.

Once you have started Stata, you can change the current working directory with the cd command. See [D] **cd** in the *Data Management Reference Manual* for full details.

Stata always displays the name of the working directory so that it is easy to tell where your graphs and datasets will be saved.

Window types

The Stata for Windows interface has two types of windows: docking and nondocking. *Docking* windows are those that have special characteristics that allow them to be used with other docking windows: they can be *linked* (sharing a window with a splitter dividing the windows) or *tabbed* (sharing a window with a tab for each). *Nondocking* windows have none of these special characteristics.

You can distinguish between docking and nondocking windows by their titlebars. Docking windows have a push-pin button in the titlebar; nondocking windows do not.

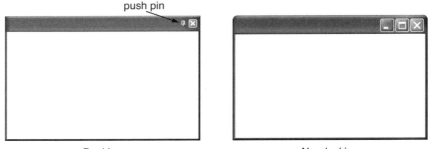

| Docking | Nondocking |

Windowing capabilities:

Docking windows
- can be dragged in and out of the main Stata window or other docking windows
- can be linked to other docking windows so that they share a window with a splitter dividing them
- can share a single window with other docking windows with a tab for each
- can be made to hide automatically when not in use
- can be made to always appear in front of the main Stata window

Nondocking windows
- cannot be dragged in and out of the main Stata window
- cannot be linked to other windows
- cannot be tabbed to another window
- cannot be made to hide automatically when not in use
- can be made to always appear in front of the main Stata window

---

When Stata is first launched, the Review, Variables, Results, and Command windows appear within the main Stata window. The Review, Variables, and Command windows are docking windows and are initially linked to each other. The Results window is a nondocking window.

Docking windows

Docking windows can be dragged into (docked) and out of (undocked) the main Stata window or other docking windows. When you drag a docking window, a transparent blue box appears in place of the window. If you drag the box over the main Stata window or another docking window, docking guides appear (see figure 1). Dragging the box over a guide previews how and where the window will appear if you release the mouse button. Releasing the mouse button when the box is over a guide (dropping) links or tabs the window as previewed, while releasing it when the box is not on a docking guide simply moves the window to that position.

Dropping the box on one of the outer docking guides links the docking window to the window under the guide. For example, figure 1 shows how you would drag the Review window over the Variables window, at which point the docking guides appear. When you move the mouse over the bottom docking guide and release the mouse button, the Review window becomes linked to the Variables window and is displayed in the lower part of the window.

When docking windows are linked, a divider (splitter) is inserted between them. The splitter allows the linked windows to be resized: increasing the size of one window decreases the size of the other. To prevent splitters from resizing linked windows, select **Prefs > General Preferences...**, click on the **Windowing** tab, and check **Lock splitters**. To undock a linked window, double-click on the window's titlebar, or drag the window away from the other window.

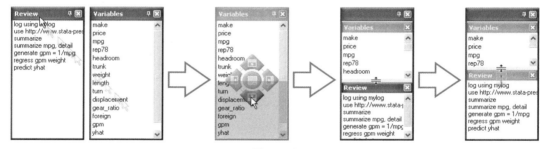

Figure 1

Dropping the box on the center docking guide (see figure 2) combines the docking window with the window under the guide. A single docking window is created with a tab for each docked window at the bottom of the new window. Clicking on a tab displays its window. To undock a tabbed window, double-click on the window's tab, or click and drag the window's tab out of the docking window. The docking window containing the tabbed windows can also be linked to other docking windows.

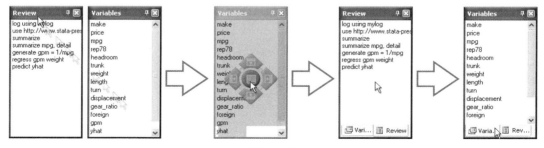

Figure 2

Docking windows, continued

You can temporarily disable the docking guides by holding down the *Ctrl* key while dragging a docking window. To permanently disable the docking guides, select **Prefs > General Preferences...**, click on the **Windowing** tab, and uncheck **Use docking guides**.

Docking windows support the Auto Hide feature, which automatically hides the docking windows when they are not in use. To enable Auto Hide for a docking window, click the push pin in the titlebar (see figure 3) so that the push pin points left. Click the push pin again to disable Auto Hide; the push pin will point down.

After you enable Auto Hide for a docking window, the window is hidden when not in use and is displayed as a tab along an edge of the main Stata window. To show the window, move the mouse pointer over the tab. To hide the window again, move the mouse pointer outside of the window. If the window has the focus (its titlebar is highlighted), it will be hidden if you click on another window. Auto Hide only works on docking windows that are docked within the main Stata window.

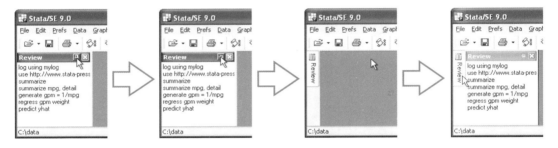

Figure 3

When you undock a docking window, you can move it in front of or behind other Stata windows, including the main Stata window. If you click on any part of the main Stata window, the main window will move to the front and completely obscure the undocked window. You can bring the undocked window back to the front by selecting it from the **Window** menu. Or you can just make undocked windows always appear in front of the main Stata window (float) by selecting **Prefs > General Preferences...**, clicking on the **Windowing** tab, and checking **Make windows float**.

Notes

In addition to dragging the window, you can dock or undock a docking window by double-clicking its titlebar. You can double-click the titlebar again to move the window back to its previously docked or undocked location.

If you enable Auto Hide for the Review and Variables windows in their default docked location, they will obscure any text at the beginning of the Command window when they are displayed. To avoid this problem, dock these two windows at the right edge of the main Stata window before enabling Auto Hide.

Nondocking windows

Nondocking windows cannot be dragged in and out of the main Stata window and do not share any of the special features of docking windows, such as linking and tabbing. Nondocking windows include the Results window, Graph window, Data Editor, Do-file Editor, and dialogs.

As with docking windows, the main Stata window, when brought to the front, can completely obscure nondocking windows that exist outside the main window. To prevent this, select **Prefs > General Preferences...**, click on the **Windowing** tab, and check **Make windows float**. This setting makes both undocked docking windows and nondocking windows always appear in front.

Notes

Making windows float is very useful if you like to run the main Stata window maximized. However, windows that are floating can obscure part of the Results window because they always appear on top of the main Stata window. You can avoid this problem by selecting **Prefs > General Preferences...**, clicking on the **Windowing** tab, and checking **Make Results window float**. This setting allows the Results window to exist outside of the main Stata window and float in front of it. You can move other windows in front or behind the Results window when it is floating.

Even though the Results window is a nondocking window, it can be linked to docked windows within the main Stata window. To do this, disable floating for the Results window if it is enabled, and then maximize the window. When the Results window is maximized, it will resize when the docked windows to which it is linked or the main Stata window are resized. However, this can affect how output is displayed. Most output in Stata is formatted for an 80-column window so resizing the docked windows or the main Stata window can decrease the available columns in the Results window. Maximizing the Results window is only practical if you are never going to resize the main Stata window or its contents. In its normal state, the Results window is not affected by the size of other windows.

Managing the user interface

If you have Stata's window settings exactly as you want them and want to prevent unintentional docking or undocking of docking windows, select **Prefs > General Preferences...**, click on the **Windowing** tab, and uncheck **Enable ability to dock, undock, pin, or tab windows**.

Stata comes with default preference sets for the user interface in the **Prefs > Manage Preferences > Load Preferences** menu. These preference sets include optimal windowing settings for running Stata maximized or pinning the Review and Variables window to the right edge of the main Stata window. You can modify any of these preference sets or create your own.

If you load a preference set from the **Prefs > Manage Preferences > Load Preferences** menu, any changes you make to the user interface will not affect the saved preference set. Changes to Stata's interface are automatically saved in a default preference set. The default preference set is loaded when Stata is started, and saved when Stata is exited. See chapter 17 for more information about managing multiple preference sets.

Notes

5 Using the Viewer

The Viewer

In this chapter, you will learn:

Stata has a Viewer in which you can
- View and print Stata help files and log files
- View and print Stata Markup and Control Language (SMCL) files
- View and print various other file types

Some Stata features invoke the Viewer:
- Viewing Stata logs
- Accessing Stata's help system
- Checking for updates to Stata

You can have more than one Viewer open at a time by
- Selecting **Window > Viewer > New Viewer**
- Clicking the **Viewer** button on the toolbar
- Middle-clicking on a link to open the link in a new Viewer (requires a 3-button mouse)
- Right-clicking on a Viewer window and selecting **Open New Viewer**
- Entering a help command from the command line

The Viewer has buttons and an edit field for navigating, entering commands, and searching

To enter the Viewer and look at a local file, such as a SMCL file or a Stata log:
- Select **File > View...**
- Enter the *filename* in the field
- Or click the **Browse...** button, and select the file you want to view
- Click **OK**

(*Continued on next page*)

The Viewer, continued

Once a log has been started, to view it in the Viewer:
- Select **File > Log > View...**
- A dialog will appear with the path and filename of the current log
- Click **OK**
- The log that appears is a snapshot of the log at the time you opened the Viewer
- Click the **Refresh** button at the top of the Viewer to see an updated snapshot of the current log

To enter the Viewer and look at a file over the Internet:
- Select **File > View...**
- Enter the URL of the file in the field
- Click **OK**

To print the contents of the Viewer:
- Right-click on the Viewer, and select **Print...**
- Or select **File > Print >** *"viewer name"*

Function of the Viewer

The Viewer is where you can see help information, view and print logs of your current and previous Stata sessions, view and print any other Stata formatted (SMCL) or plain text (ASCII) file, add new commands to Stata from the Internet, and install the latest official updates to Stata. You can either click the **Viewer** button or select **Window > Viewer > New Viewer**, and the Stata Viewer will open and provide you with links that allow you to perform a number of tasks.

```
Viewer (#1) [Advice on Help]                                          🖈 ☒

  [ Back ]  [ Refresh ]  [ Search ]  [ Help! ]  [ Contents ][What's New][ News ]

  Command: help viewer                                                    🔍

  ────────────────────────────────────────────────────────────────────
  Using the Viewer                      manual: [GSM] 5 Using the Viewer
                                                [GSU] 5 Using the Viewer
                                                [GSW] 5 Using the Viewer
  ────────────────────────────────────────────────────────────────────

      In the Viewer, you can
      ──────────────────────────────────────────────────────────

      see help for contents or help for any Stata command

      search help files, documentation, and FAQs    (advice on using
            search)

      find and install SJ, STB, and user-written programs from the net

      review, manage, and uninstall user-written programs

      check for and optionally install official updates

      view your logs or any file
```

In the Viewer, you can

- View the Contents of the help system
- View the help file for any Stata command
- Search Stata's documentation and FAQs by *keywords*
- Search net resources by *keywords*
- Obtain advice on how to use `search`
- Find and install *Stata Journal*, STB, and user-written programs
- Review, manage, and uninstall user-written programs
- Check for and optionally install official Stata updates
- View logs and other SMCL and ASCII files
- Launch a browser
- See the latest news from *http://www.stata.com/*

Function of the Viewer, continued

The Viewer is part browser in that you can click on links (shown in blue text) to execute commands that the Viewer understands, display another help file in the Viewer, or even update your Stata. When you move the mouse pointer over a link, the status bar at the bottom of the Viewer shows the action associated with that link. If the action of a link is to show help on `logistic`, left-clicking that link will show the help file in the Viewer.

You can have more than one Viewer open at a time by selecting **Window > Viewer > New Viewer**, clicking the Viewer button on the toolbar, middle-clicking on a link, or entering a help command from the command line. Right-clicking on a Viewer window also allows you to open a new Viewer, copy text, edit the preferences, and print the contents of the window.

Stata allows you to print Viewer windows from the command line. Each Viewer is given a unique name (which is displayed within parentheses in the window's titlebar) to help identify it from the command line. For example, the name of a Viewer with the window title *Viewer (#1) [help regress]* is **#1**.

To bring a Viewer to the front of all other Viewers, select **Window > Viewer**, and choose a Viewer from the list there. Selecting **Close All Viewers** closes all open Viewer windows.

The Viewer is a docking window (see chapter 4). A Viewer that is docked with the main Stata window when Stata is exited will be opened automatically the next time Stata is started. To also make undocked Viewers automatically open, select **Prefs > General Preferences...**, click the **Windowing** tab, and check **Make undocked Viewers persistent**.

A Viewer that is automatically opened initially displays the Viewer help file. You can make it display the same contents it displayed when it was last open by selecting **Prefs > General Preferences...**, clicking on the **Windowing** tab, and checking **Persistent Viewers retain topics**.

Viewer buttons

Seven buttons appear at the top of the Viewer:

Back returns you to the previous contents viewed in the Viewer.

Refresh reloads the current contents of the Viewer.

Search allows you to perform a keyword search for help files, FAQs, *Stata Journal* articles and programs, STB articles and programs, and user-written programs.

Help! gives options for using the Viewer. It gives links, advice, and detailed instructions for using the Viewer. Especially valuable are its advice and examples for using **Search** and **Help**. Take a look at it.

Contents contains a table of contents for all the Stata help files, arranged by subject.

What's New describes the new features of Stata in the current release.

News lists recent news and information of interest to Stata users.

Viewing local files, including SMCL files

You can use the Viewer to view a variety of file types, including Stata Markup and Control Language (SMCL) files. This is a language similar to HTML that contains directives specifying how the text is to be displayed; see [P] **smcl** for details.

To open a file and view its contents, simply select **File > View...**, and you will be presented with a dialog.

```
Choose File to View

File or URL:
[                                    ]

[    OK    ]   [  Cancel  ]   [ Browse... ]
```

Type in the name of the file that you wish to view, and click **OK**. There is also a **Browse...** button that will allow you to search your hard drive for a file. For example, suppose that you wish to view a SMCL log file, named `myfile.smcl`, which you have stored in `C:\data`.

1. *Select* **File > View...**.

2. *Click the* **Browse...** *button.*

3. *Navigate the hard drive until you find the file.*

4. *Click once on the file.*

5. *Click* **Open**.

6. *Click* **OK**.

The file will now be displayed in the Viewer.

This is somewhat different from when you want to view the current log. In that case, you select **File > Log > View...**, and the usual dialog will appear, but with the path and filename of the current log already in the field. Simply click **OK**, and the log will appear in the Viewer. See chapter 16 for more details.

Note

If you log output, your logs, by default, will be written in SMCL. You must use the Viewer to view or print your SMCL logs because other programs do not understand SMCL. In addition, if you need to translate files containing SMCL to other formats, you can select **File > Log > Translate...** (see [R] **translate**). You can change the default log format; see [R] **log** for more information.

Viewing Internet files

If you want to look at a file over the Internet, the process is very similar to viewing a local text file, only instead of using the **Browse...** button, you simply type in the URL of the file that you want to see, such as *http://www.stata.com/man/readme.smcl*. Please note, however, that you do not want to just provide the URL for any web page. Web pages are written in HTML and are not translated by the Viewer. If you were to use the Viewer to look at a file over the Internet that was written in HTML, you would see the plain text HTML code of the document. You can launch your browser to view an HTML file by clicking on the browser link from the **Help!** screen or by typing `browse` *URL* in the **Command** field at the top of the Viewer. Also note that you must provide the complete path of the file that you wish to view.

Navigating the Viewer

In addition to using the window scrollbar to navigate the Viewer window, you also can use the up/down cursor keys and *Page Up/Page Down* keys to do the same. Pressing the up/down cursor keys scrolls the window a line at a time. Pressing the *Page Up/Page Down* keys scrolls the window a page at a time.

Printing

To print the contents of the Viewer, right-click on the window, and select **Print...**. You may also select **File > Print >** *"viewer name"* or click on the arrow on the **Print** toolbar button to select from a menu of open windows to print.

Right-clicking on the Viewer window

Right-clicking on the Viewer window displays a contextual menu from which you can select

- **Back** to return to the previous contents of the Viewer.
- **Open New Viewer** to open a new Viewer.
- **Open Link in New Viewer** to open a link in a new Viewer.
- **Copy** to copy text to the Clipboard.
- **Copy Table** to copy a table to the Clipboard in a tab-delimited format.
- **Copy Table as HTML** to copy a table to the Clipboard in HTML format.
- **Find in Viewer** to find text in the Viewer.
- **Preferences...** to edit the preferences for the Viewer window.
- **Font...** to change the font for the window.
- **Print...** to print the contents of the Viewer window.

Commands in the Viewer

Everything you can do in the Viewer by clicking on links and buttons can also be done by issuing commands in the **Command** field at the top of the window or on the Stata command line. Some of the commands that can be issued in the Viewer are:

1. *Obtaining help*

    Type `help contents` to view the contents of Stata's help system.

    Type `help` *commandname* to view the help file for a Stata command.

2. *Searching*

    Type `search` *keyword* to search documentation and FAQs on a topic.

    Type `search` *keyword*`, net` to search net resources on a topic.

3. *User-written programs*

    Type `net from http://www.stata.com/` to find and install *Stata Journal*, STB, and user-written programs from the net.

    Type `ado` to review user-written programs.

4. *Updating*

    Type `update` to check your current Stata version.

    Type `update query` to check for new official Stata update releases.

    Type `update all` to update your Stata.

5. *Viewing files in the Viewer*

    Type `view` *filename*`.smcl` to view SMCL files.

    Type `view` *filename*`.txt` to view ASCII files.

    Type `view` *filename*`.log` to view ASCII log files.

6. *Viewing files in the Results window*

    Type `type` *filename*`.smcl` to view SMCL files.

    Type `type` *filename*`.txt` to view ASCII files.

    Type `type` *filename*`.log` to view ASCII log files.

7. *Launching your browser to view an HTML file*

    Type `browse` *URL* to launch your browser.

8. *Keeping informed*

    Type `news` to see the latest news from *http://www.stata.com*.

6 Help

Online help

In this chapter, you will learn:

Stata has a comprehensive **Help** system.

When you open the **Help** menu, you get a menu from which you can
- See the help table of contents
- Search for help entries on a topic
- Get help for a Stata command
- List new features in Stata
- Read the latest Stata news
- Install the latest official updates from the Internet
- Install user-written, *Stata Journal*, or *Stata Technical Bulletin* programs for Stata from the Internet
- Go straight to important points on the Stata web site

Selecting **Help > Search...** allows you to
- **Search documentation and FAQs** by *keywords*
- **Search net resources** by *keywords*

Choosing **Search documentation and FAQs** from the **Search...** dialog opens a Viewer containing
- Links that will take you to the help files for the appropriate Stata commands
- References to the topic in the Stata Documentation
- References to the topic in the *Stata Journal* and in the *Stata Technical Bulletin*
- Links to FAQs on Stata's web site dealing with the topic

Choosing **Search net resources** from the **Search...** dialog opens a Viewer containing
- Links that will take you to files on the Internet containing *Stata Journal*, STB, and user-written programs related to your *keywords*

(*Continued on next page*)

Help, continued

Multiple topic words are allowed with **Search**
  Adding topic words will narrow the search; for example,
  - Enter `regression residuals`
  Use proper English and statistical terminology when doing a **Search**.
  For example,
  - `t test` with **Search...** is correct
  - `ttest` with **Stata command...** is correct

Selecting **Help > Contents** gives a listing of Stata's help
  table of contents.
  - You may choose from the links on this page to view help
    for a particular command, or
  - you may enter the full name of a Stata command in
    the edit field at the top of the Viewer

The command help files also contain links.
  - Click on a link, and you will go to another help file
  - Click on a dialog link, and you will see the dialog

When you have stepped through a series of help files,
  - Click the **Back** button to go back to the previous help file

The Help system

Stata's help system provides a wealth of information to help you learn and use Stata. To find out which Stata command will perform the statistical or data-management task you would like to do, you should generally follow these steps:

1. Select **Help > Search...**, choose **Search documentation and FAQs**, and enter the topic.

2. Scan the **Search** results, and click on the appropriate command name link to open its help file.

3. With the help file open, click on the Command window and enter the command, or click on one of the dialog links to open a dialog for the command.

4. If the first help file you went to is not what you wanted, either look at the end of the help file for links to take you to related help files, or click the **Back** button to go back to the previous document and go from there to other help files.

5. If, at any time, you want to begin again with a new **Search**, click the **Search** button, or select **Help > Search...**.

6. If your **Search documentation and FAQs** search returned no results, we suggest that you look for *Stata Journal*, STB, and user-written programs on the Internet (materials available via Stata's net command) by selecting **Search net resources** and entering your topic again.

   If you select **Search documentation and FAQs**, Stata searches Stata's keyword database. If you select **Search net resources**, Stata searches for *Stata Journal*, STB, and user-written programs available for free download on the Internet; see chapter 19 for more information.

7. Alternatively, you can select **Search all**. **Search all** is like choosing both **Search documentation and FAQs** and **Search net resources**.

The Help system, continued

Let's illustrate the **Help** system with an example. Selecting **Help > Search...** gives you two options for finding information about Stata:

- **Search documentation and FAQs** allows you to search by keyword for help files for specific Stata commands, references to the Stata Documentation, references to the *Stata Journal* and *Stata Technical Bulletin*, and links to FAQs on Stata's web site.

- **Search net resources** allows you to search from Internet resources beyond those maintained by Stata.

Say that you want to search for information resources containing the word *data*. You would follow these steps:

1. *Select* **Help > Search...**

2. *Choose* **Search documentation and FAQs**.

3. *Enter* data, *and click* **OK** *or press Enter.*

The Help system, continued

4. *Stata will now search for all references to "data" among the Stata commands, the Reference manuals, the User's Guide, the Stata Journal, the Stata Technical Bulletin, and the FAQs on Stata's web site.* Here is the **Search** result:

```
Viewer (#1) [search data]                                              ⱷ ⊠

  [  Back  ] [  Refresh  ] [  Search  ] [  Help!  ] [  Contents ][What's New] [  News  ]

  Command: search data                                                    🔍

  ─────────────────────────────────────────────────────────────────────── ⌃
  search for data                                    (manual:  [R] search)

          Keywords:  data
          Search:    (1) Official help files, FAQs, Examples, SJs, and STBs

  Search of official help files, FAQs, Examples, SJs, and STBs

  GS         . . . . . . . . . . . . . . . . . . . . . . Getting Started manual

  [GSM]      17 . . . . . . . . . . . . . . . Setting font and window preferences
             (help macfonts)

  [GSW]      17 . . . . . . . . . . . . . . . Setting font and window preferences
             (help winfonts)

  [GSU]      17 . . . . . . . . . . . . . . . Setting font and window preferences
             (help unixfonts)

  [U]        Chapter 1.2.1 . . . . . . . . . . . . . . . . . . Sample datasets
             (help dta_contents, sysuse, webuse, whatsnew8to9)

  [U]        Chapter 12 . . . . . . . . . . . . . . . . . . . . . . . . . Data  ⌄
             (help datatypes, label, missing, notes, format)

  ────────────────────────────────────────────────────────────────────────
```

The Help system, continued

5. *Scroll down until you see* [D] describe.

describe is a Stata command that will "Describe data in memory or in file". [D] means that the describe command is documented in the *Stata Data Management Reference Manual*. The "help describe" means that there is an online help file for it.

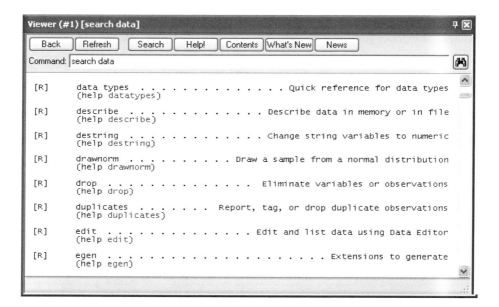

The Help system, continued

6. *Click on the link* "describe" *in* "help describe".
   Blue words in Stata are hypertext links. When you click on a link, Stata displays the corresponding help file, web page, etc. Clicking on the blue "describe" link displays the help file for the `describe` command.

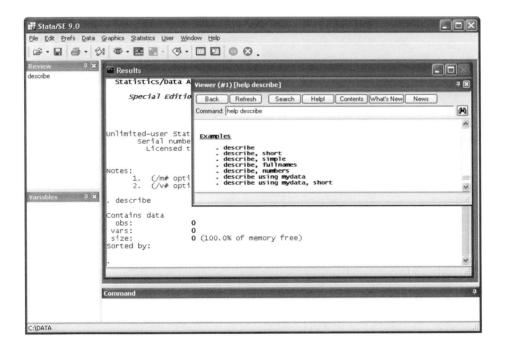

The Help system, continued

7. *Look at the help file for* describe.

Help files for Stata commands contain (1) the command's syntax (i.e., the rules for typing it), (2) links to dialogs, (3) a description of the command, (4) options, (5) examples, and (6) references to related commands. As you go through this manual, you will learn all about these things. For now, just scroll down to the examples. The examples show that it is valid to just type describe without anything else.

8. *Click on the Command window, type* describe, *and press Enter.*

You may have to resize the Viewer so that you can see the bottom of the Results window. You can see that describe tells us that we have no observations.

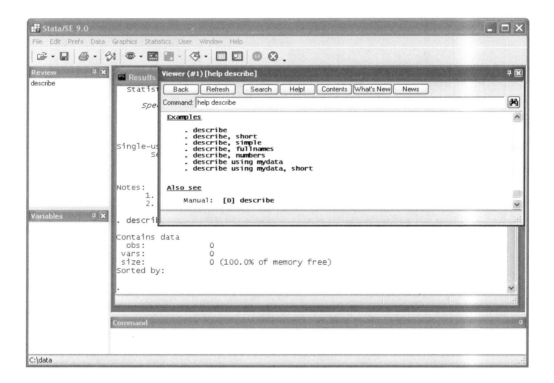

Searching

Search is designed to help you find information about statistics, graphics, data management, and programming features in Stata. When entering topics for the search, use appropriate terms from statistics, etc. For example, you could enter `Mann-Whitney`. Multiple topic words are allowed, e.g., `regression residuals`. For advice on using **Search**, click the **Help!** button, and click on **advice** in (**advice on using search**).

When you are using **Search**, use proper English and proper statistical terminology. Don't choose **Search...** and enter `ttest`. The term "ttest" is not proper statistical terminology. It is, however, the name of a Stata command. If you already know this and want to go directly to the help file for the `ttest` command, select **Help > Stata command...**, and type `ttest`. You can also type `help ttest` in the edit field at the top of the Viewer and press *Enter*.

Help needs to distinguish between topics and Stata commands since some of the names of Stata commands are general topics. For example, `logistic` is a Stata command. If you choose **Stata command...** and enter `logistic`, you will go right to the help file for the command. But, if you choose **Search...** and enter `logistic`, you will get search results listing the many Stata commands that relate to logistic regression.

Contents

If you click the **Contents** button, you will see several help categories.

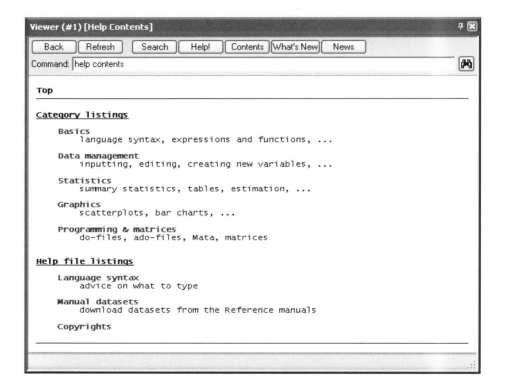

Click on a link to display more information about that category. For example, if you click on **Statistics**, more information about statistics will be displayed.

help and search commands

Notes

1. *You can also access Stata's help system from the Command window.*
   When you do so, the output can appear either in the Results window or in the Viewer.

2. *Typing* `search` *topic*
   in the Command window produces the same output as selecting **Help > Search...**, choosing **Search documentation and FAQs**, and entering *topic*. However, the output will appear in the Results window.

3. *Typing* `search` *topic*`, net`
   in the Command window produces the same output as selecting **Help > Search...**, choosing **Search net resources**, and entering *topic*. However, the output will appear in the Results window.

4. *Typing* `search` *topic*`, all`
   in the Command window produces the same output as selecting **Help > Search...**, choosing **Search all**, and entering *topic*. However, the output will appear in the Results window.

5. *Typing* `help` *commandname*
   is equivalent to selecting **Help > Stata command...** and entering *commandname*. The help file will appear in Stata's Viewer.

6. *Typing* `chelp` *commandname*
   gives the same output as selecting **Help > Stata command...** and entering *commandname*, except that the output will appear in the Results window.

7. See [U] **4 Stata's online help and search facilities** and [U] **4.8 search: All the details** in the *User's Guide* for more information about these command language versions of the **Help** system. The `search` command, in particular, has a few capabilities (such as author searches) that we have not demonstrated here.

The Stata Reference manuals and User's Guide

Notations such as [R] **ci**, [R] **regress**, and [R] **ttest** in the **Search** results and help files are references to the 3-volume *Stata Base Reference Manual*. You may also see things like [P] **#delimit**, which is a reference to the *Stata Programming Reference Manual*, and [U] **9 The Break key**, which is a reference to the *Stata User's Guide*. For a complete list of manuals and their shorthand notations, see *Cross-Referencing the Documentation*, which immediately follows the table of contents in this manual. Why bother to look in the manual when help is available in Stata? The manual has more information. Although Stata's interactive help is extensive, it contains only about one-tenth of the information contained in the Reference manuals.

The Stata Reference manuals are each arranged like an encyclopedia, alphabetically, and each has its own index. The *User's Guide* also has an index. The *Quick Reference and Index* contains a combined index for the *User's Guide* and all the Reference manuals. This combined index is a good place to start when you are looking for information about a command.

Entries have names like **collapse**, **egen**, . . . , **summarize**, which are generally themselves Stata commands.

For advice on how to use the Reference manuals, see chapter 18, or see [U] **1.1 Getting Started with Stata**.

The Stata Journal and the Stata Technical Bulletin

The **Search documentation and FAQs** facility searches all the Stata source materials, including the online help, the *User's Guide*, the Reference manuals, this manual, the *Stata Journal*, the *Stata Technical Bulletin* (STB), and the FAQs on Stata's web site.

The **Search net resources** facility searches all materials available via Stata's **net** command; see [R] **net**. These materials include the *Stata Journal*, STB, and user-written additions to Stata available on the Internet.

The *Stata Journal*, which began in 2001, is a quarterly journal containing articles about statistics, data analysis, teaching methods, and effective use of Stata's language. The journal publishes reviewed papers together with shorter notes and comments, regular columns, book reviews, and other material of interest to researchers applying statistics in a variety of disciplines. See *http://www.stata-journal.com* for additional information.

The predecessor to the *Stata Journal* is the *Stata Technical Bulletin* (STB). Even though the STB is no longer published, past issues contain articles and programs that may interest you. See *http://www.stata.com/bookstore/stbj.html* for the table of contents of past issues, and see the STB FAQ at *http://www.stata.com/support/faqs/res/stb.html* for detailed information on the STB.

Associated with each issue of both the *Stata Journal* and the STB are the programs and datasets described in the journals. These programs and datasets are made available for download and installation over the Internet, not only to subscribers, but to all Stata users. See [R] **net** and [R] **sj** for more information.

Since the *Stata Journal* and the STB have had a number of articles dealing with correlations, if you select **Help > Search...** and choose **Search documentation and FAQs**, and enter `correlations`, you will see some of these references:

(*Continued on next page*)

The Stata Journal and the Stata Technical Bulletin, continued

```
Viewer (#1) [search correlations]                                        ⊞ ☒
 [ Back ]  [ Refresh ]  [ Search ]  [ Help! ] [ Contents ][What's New][ News ]
 Command: search correlations                                              [🔍]
 SJ-2-1    st0009 . . . . . . . . . . . . . . . . . . . . Transfer functions  ▲
           . . . . . . . . . . . . . . . . . . . . . . . . . . . A. McDowell
           Q1/02   SJ 2(1):71--85                          (no commands)
           demonstrates the estimation of a transfer function using
           Stata's arima command

 SJ-1-1    st0004  Residual diagnostics for cross-section time series reg. models
           (help xttest2, xttest3 if installed) . . . . . . . . . . . C. F. Baum
           Q4/01   SJ 1(1):101--104
           commands for testing groupwise heteroskedasticity and cross-
           sectional correlation after xtreg, fe and xtgls

 STB-61    sg164 . . . . . . . . . Specification tests for linear panel data models
           (help xttest1 if installed) . . . . . W. Sosa-Escudero and A. K. Bera
           5/01    pp.18--21; STB Reprints Vol 10, pp.307--311
           extension of xttest0 that computes seven specification tests
           for balanced error component models

 STB-61    snp15.3 . . . . . . . . . . . . . . . . . . . . . . Update to somersd
           (help somersd if installed) . . . . . . . . . . . . . . . . R. Newson
           5/01    p.22; STB Reprints Vol 10, p.324
           updated for compatibility with Stata 7

 STB-59    sg159 . . . . . . . . . . . . Confidence intervals for correlations
           (help ci2 if installed) . . . . . . . . . . . . . . . . . P. T. Seed
           1/01    pp.27--28; STB Reprints Vol 10, pp.267--269
           enhancement of ci and cii that provide confidence intervals
           for Pearson's product moment correlation and Spearman's
           rank correlation                                                  ▼
```

SJ-2-1 refers to Volume 2 Number 1 of the *Stata Journal*. Abbreviations like st0004, st0009, etc., are the way the *Stata Journal* refers to the articles that have appeared in it.

STB-59 refers to the fifty-ninth issue of the STB. Every six issues make a year's worth of STBs, and they are bound and republished in a volume called *Stata Technical Bulletin Reprints*. STB-59 could thus be found in *STB Reprints*, Volume 10. Abbreviations like sg159, snp15.3, etc., are the way the STB refers to the articles that have appeared in it.

In the screenshot above, you will notice that article st0004 in SJ-1-1 has two commands, `xttest2` and `xttest3`, associated with it. You can install these commands using Stata's Viewer. You may also install commands from STB articles via the Internet. See chapter 19 for more information.

Links to other sites where you can freely download programs and datasets for Stata can be found on the Stata web site; see *http://www.stata.com/links/*. See chapter 19 for more details on how to install this software. Also see [R] **ssc** for information on a convenient interface to resources available from the Statistical Software Components (SSC) archive.

We recommend that all users subscribe to the *Stata Journal*. See [U] **3.5 The Stata Journal and the Stata Technical Bulletin** for more information.

7 Loading and saving data

How to load your dataset from disk and save it to disk

In this chapter, you will learn:

To load a Stata data file:
- Select **File > Open...**
- Select **File > Open Recent >** *filename*
- Or type use *filename*
- Or click the **Open** button

When loading a dataset, the dataset currently in memory is discarded.
Either save the current dataset first or allow it to be discarded:
- Choosing a file after selecting **File > Open...** will prompt you before discarding it
- Or type use *filename*, clear
- Or type clear and then use *filename*

To save a new dataset (or an old dataset under a new name):
- Select **File > Save As...**
- Or type save *filename* in the Command window

To save a dataset for use with Stata 7:
- Select **File > Save As...**, and select
  Stata 7 Data (*.dta) from the **Save as type** list
- Or type saveold *filename* in the Command window

To resave a dataset that has been changed (overwriting the original data file):
- Select **File > Save**
- Or click the **Save** button
- Or type save *filename*, replace
- Or simply type save, replace

Stata data files are named *filename*.dta. When saving or loading data files, the extension .dta is automatically added when an extension is not specified.

Datasets used in examples in the Stata manuals are available online.
Visit *http://www.stata-press.com/data/*

Important note: Changes are not permanent until you save them. You work with a copy of the dataset in memory, not with the data file itself.

Comment about resaving datasets: There is no way to recover your original data file once you have done a **File > Save** or save, replace. With important datasets, you may want either to keep a backup copy of your original *filename*.dta or to save the changed dataset under a new name.

Increasing memory for data: Stata loads your data into memory before analyzing it. By default, Stata/SE uses 10 megabytes of memory for data, and Intercooled Stata uses 1 megabyte of memory. Depending on the size of your datasets, you may wish to adjust the amount of memory Stata allocates for data. For full details, see [GSW] **C. Setting the size of memory**.

Notes

8 Using the Data Editor

The Data Editor

In this chapter, you will learn:

To enter the editor:
- Click the **Data Editor** button
- Or type `edit` and press *Enter* in the Command window

The editor is like a spreadsheet.
- Columns correspond to variables and rows to observations.
- You navigate by clicking on a cell or by using the arrow keys

You can copy and paste data between Stata's editor and other spreadsheets:
- Highlight the data that you wish to copy in either Stata or the other spreadsheet, right-click on the window, and then select **Copy**
- After copying the data, you can paste it into Stata's editor or into another spreadsheet by selecting the top left cell of the area to which you wish to paste, right-clicking on the window, and selecting **Paste**

You modify or enter data by
- Choosing the cell, typing the value, and pressing *Enter* or *Tab*
- *Enter* highlights the next cell in the column
- *Tab* highlights the next cell to the right in the row (until you get to the last filled-in column, and then it highlights the first column again)

To input data variable by variable:
- Click on the top cell in the first empty column
- Type the value
- Press *Enter* to highlight the next cell in that column

To input data observation-by-observation:
- Click on the first cell in the first empty row
- Type the value
- Press *Tab* to highlight the next cell to the right
- After the first observation's values have been entered, click on the second cell in the first column
- Type the values for the second observation; press *Tab* to highlight the next cell to the right
- At the end of the second observation (and all subsequent ones), *Tab* will automatically highlight the first column again

---

(Continued on next page)

The Data Editor, continued

Numeric data and string data (i.e., data consisting of characters) are entered the same way.
- You do not have to type double quotes around strings

Missing values for numeric variables are recorded as '.' (i.e., a period). To enter a missing numeric value:
- Press *Enter* or *Tab*
- Or type '.' and press *Enter* or *Tab*

Stata supports different types of missing values if needed; see [U] **12.2.1 Missing values** for more information

Missing values for string variables are just empty strings (i.e., nothing). To enter a missing string value:
- Just press *Enter* or *Tab*

The editor initially names variables var1, var2, ... To rename a variable:
- Double-click anywhere in the variable's column

This opens the **Variable Properties** dialog:
- Type the new name of the variable, and click **OK**

A variable name must be 1 to 32 characters long.
- The characters can be letters: A – Z, a – z
- Or digits: 0 – 9
- Or underscores: _
- But no spaces or other characters

Example: My_Name1

The first character must be a letter or an underscore, but using an underscore to begin your names is not recommended.

Stata is case-sensitive.
- Myvar, myvar, and MYVAR are different names

The Data Editor, continued

You can select the variables that appear in the editor:

Type in the Command window:

- edit make   Selects the single variable make
- edit make mpg   Selects the variables make and mpg

You can include any number of variables.

You can restrict the observations that appear in the editor:

Type in the Command window:

- edit in 1   Uses only the first observation
- edit in 2   Uses only the second observation
- edit in -2   Uses only the second from the last observation
- edit in -1   Uses only the last observation
- edit in l   (i.e., ℓ)   Also uses only the last observation

To restrict to a series of observations:

- edit in 1/9   Uses observations 1 through 9
- edit in 2/-2   Uses observations 2 through second from the last

To restrict to those observations satisfying a mathematical expression:

- edit if *exp*   Uses observations for which *exp* is true
- edit if mpg>20   Uses observations for which mpg>20
- edit if mpg==20   Uses observations for which mpg equals 20
- edit if missing(rep78)   Uses observations for which rep78 equals missing

You can combine in and if (the order does not matter):

- edit in 1/9 if mpg>=25
- edit if price<6000 in 5/-1

You can select variables and restrict observations at the same time:

- edit make in 5/-5
- edit make mpg if mpg>=25
- edit make mpg if missing(rep78)
- edit make mpg in 1/9 if mpg>=25

(Continued on next page)

The Data Editor, continued

Restricting the editor to certain variables and observations lessens the chance of making a serious mistake. For example, if you need to change mpg when it is missing, you can limit the data's exposure by typing
- edit make mpg if missing(mpg)

Note: For making systematic global changes to your data, the replace command may be more appropriate.
- See chapter 12

To delete variables and observations:
- Click the **Delete...** button; see the next page

Note: For deleting groups of observations or variables, the drop command is more appropriate.
- See chapter 13

The Data Editor can be used to view data.
To enter the editor in browse mode:
- Click the **Data Browser** button
- Or type browse in the Command window

The browse mode is useful since you cannot accidentally change your data. Use browse, not edit, when you just want to look.

You can also select variables and restrict observations with browse.
Examples:
- browse make mpg
- browse in 1/20
- browse if missing(rep78)
- browse make mpg rep78 in 5/-5 if mpg>20

In later chapters, you will learn that what is true for edit and browse is true for almost all Stata commands: You type the command, optionally followed by a variable list, optionally followed by an if, optionally followed by an in.

Buttons on the Data Editor

The editor has seven buttons:

Preserve

If you make changes to your data in the Data Editor and are satisfied with them (and plan to stay in the editor to make additional changes), you can update the backup copy by clicking **Preserve**.

Restore

Stata automatically makes a backup copy of your data when you enter the editor.

If you want to cancel the changes that you made and restore the backup copy, click **Restore**.

Sort

Sort sorts the observations of the current variable in ascending order.

<<

The << button shifts the current variable to be the first variable.

>>

The >> button shifts the current variable to be the last variable.

Hide

Hide hides the current variable.
The variable still exists; the editor just stops displaying it.

Delete...

Delete... opens a dialog that allows you to
- Delete the current variable
- Delete the current observation
- Delete all observations throughout the dataset that have the same value for the current variable as the current observation

Inputting data

Again let's illustrate using an example. Suppose that you have the following dataset:

Make	Price	MPG	Weight	Gear Ratio
VW Rabbit	4697	25	1930	3.78
Olds 98	8814	21	4060	2.41
Chev. Monza	3667		2750	2.73
AMC Concord	4099	22	2930	3.58
Datsun 510	5079	24	2280	3.54
	5189	20	3280	2.93
Datsun 810	8129	21	2750	3.55

Note that you do not know MPG for the third car or the make of the sixth.

You will enter this dataset using the Data Editor.

1. *Open the editor:*
   Click the **Data Editor** button, or type `edit` in the Command window.

Inputting data, continued

2. *Enter the data.*
   Data can be entered variable by variable or observation by observation. Columns correspond to variables and rows to observations.

3. *When entering data observation by observation, press Tab after each value.*
   To input data observation by observation, start in the top cell of the first column. Type the make VW Rabbit, and press *Tab* to highlight the next cell to the right. Do not press *Enter* since it highlights the cell below.

   Now enter the price 4697, and press *Tab*. Continue until you have entered all the values for the first observation. Now click on the second cell in the first column, and enter the values for the second observation using the *Tab* key.

4. *The Tab key is smart.*
   After the first observation has been entered, Stata knows how many variables you have. So, at the end of the second observation (and all subsequent observations), *Tab* will automatically take you back to the first column.

5. *When entering data variable by variable, press Enter after each value.*
   To enter data variable by variable, click on the top cell in the first empty column. Type the values for the variable, and press *Enter* after each one.

Inputting data, continued

Things to know about entering data

1. *Quotes around strings are unnecessary.*
   For other Stata commands, you will learn that you must type double quotes (") around strings. You can use double quotes in the editor, too, but you do not have to bother.

2. *Strings and value labels are color coded.*
   To help distinguish between the different types of variables in the editor, string values are displayed in red, value labels (see chapter 10) are displayed in blue, and all other values are displayed in black. You can change the colors for strings and value labels by right-clicking on the editor window, and selecting **Preferences...**.

3. *A period ('.') represents Stata's system missing numeric value.*

4. *Just press Tab or Enter to input a system missing numeric value.*
   MPG for the third observation is missing. To enter the missing value, just press *Tab* (or *Enter*), or type '.'.

5. *Just press Tab or Enter to input a missing value for a string variable.*
   You do not know the make of the sixth car, so just press *Tab* (or *Enter*), and there will be an empty string (i.e., nothing) for the value of make for this observation.

6. *Stata will not allow empty columns or rows in the middle of your dataset.*
   Whenever you enter new variables or observations, always begin in the first empty column or row. If you skip over some columns or rows, Stata will fill in the intervening columns or rows with missing values.

7. *When you see, for example,* var3[4]=
   This corresponds to the current cell that is highlighted. var3 is the default name of the third variable, and [4] indicates the fourth observation.

	var1	var2	var3	var4	var5
1	VW Rabbit	4697	25	1930	3.78
2	Olds 98	8814	21	4060	2.41
3	Chev. Monza	3667	.	2750	2.73
4	AMC Concord	4099	22	2930	3.58
5	Datsun 510	5079	24	2280	3.54
6		5189	20	3280	2.93
7	Datsun 810	8129	21	2750	3.55

Data Editor — Preserve · Restore · Sort · << · >> · Hide · Delete... — var3[4] = 22

Renaming variables

1. *The editor initially names variables* var1, var2, ..., var5 *when it creates them.*

2. *To rename a variable, double-click anywhere in the variable's column.*
   Double-clicking on any cell in the variable's column opens the **Variable Properties** dialog. Type the new name of the variable, and click **OK**. (You can also enter a variable label; see chapter 10. You can also change the display format of the variable; see [U] **12.5 Formats: controlling how data are displayed**.)

   Name the first variable make, the second price, the third mpg, the fourth weight, and the fifth gear_ratio. Just before you double-clicked and renamed var5 to gear_ratio, your screen looked like this:

	make	price	mpg	weight	var5
1	VW Rabbit	4697	25	1930	3.78
2	Olds 98	8814	21	4060	2.41
3	Chev. Monza	3667	.	2750	2.73
4	AMC Concord	4099	22	2930	3.58
5	Datsun 510	5079	24	2280	3.54
6		5189	20	3280	2.93
7	Datsun 810	8129	21	2750	3.55

Data Editor — Preserve | Restore | Sort | << | >> | Hide | Delete... — var5[3] = 2.73

Rules for variable names

1. *Stata is case-sensitive.*
   Make, make, and MAKE are all different names to Stata. If you had named your variables Make, Price, MPG, etc., you would have to type them correctly capitalized in the future. Using all lowercase letters is easier.

2. *A variable name must be 1 to 32 characters long.*

3. *The characters can be letters (A–Z, a–z), digits (0–9), or underscores (_).*

4. *Spaces or other characters are not allowed.*

5. *The first character of a variable name must be a letter or an underscore.*
   Using an underscore to begin your names is not recommended since Stata's built-in variables begin with an underscore.

Copying and pasting data

1. *Select the data that you wish to copy.*

   a. Click once on a variable name to select the entire column.

   b. Click once on an observation number to select the entire row.

   c. Click and drag the mouse to select a range of cells.

   As an example, you will copy and paste an observation from Stata's editor into itself. Highlight the observation by clicking on the observation number.

2. *Copy the data to the Clipboard.*
   Right-click on the selection, and select **Copy**.

Copying and pasting data, continued

3. *Paste the data from the Clipboard.*

   a. Click on the top left cell of the area to which you wish to paste.

   b. Right-click on the cell, and select **Paste**.

Changing data

To change data using the Data Editor, follow these steps:

1. *Load the* `auto.dta` *dataset that we provide with Stata:*
   Select **File > Open...**, and then select `auto.dta` from the `C:\Program Files\Stata9` directory (or from wherever Stata is installed). You cannot be using the Data Editor and open an existing dataset.

2. *Enter the editor:*
   Click the **Data Editor** button, or type `edit` in the Command window.

3. *The* **Sort** *button sorts the observations in ascending order of the current variable.*
   Suppose that you want to look at the lowest and highest priced cars. Click once anywhere on the `price` column, and then click **Sort**. Scroll down to the bottom of the dataset to see the highest priced cars.

4. *The shift button* << (>>) *shifts the current variable to be the first variable* (*last variable*).
   Say that you want to find the lightest and heaviest cars. Unless you have a big monitor, you will not be able to view both `make` and `weight` together. So, scroll right, click on the `weight` column, and then click <<. Now, scroll left, select `weight`, and click **Sort**. Select `make` and click <<.

		make	weight	price	mpg	
	1	Honda Civic	1,760	4,499	28	
	2	Ford Fiesta	1,800	4,389	28	
	3	Plym. Champ	1,800	4,425	34	
	4	Renault Le Car	1,830	3,895	26	
	5	VW Rabbit	1,930	4,697	25	
	6	Mazda GLC	1,980	3,995	30	
	7	VW Scirocco	1,990	6,850	25	
	8	Datsun 210	2,020	4,589	35	
	9	VW Diesel	2,040	5,397	41	
	10	Subaru	2,050	3,798	35	
	11	Audi Fox	2,070	6,295	23	
	12	Chev. Chevette	2,110	3,299	29	
	13	Dodge Colt	2,120	3,984	30	
	14	Fiat Strada	2,130	4,296	21	
	15	VW Dasher	2,160	7,140	23	
	16	Toyota Corolla	2,200	3,748	31	

Changing data, continued

5. *The **Hide** button hides the current variable.*
   The effect of **Hide** is only cosmetic. The variable still exists; it is simply not displayed. Click on the rep78 column, and click **Hide**.

6. *To change a value, click on the cell, type the new value, and press Enter or Tab.*

7. ***Delete...** allows you to delete one variable, one observation, or all observations that have a certain value.* Select the trunk variable. Click **Delete...**, and click **OK**. You have now dropped trunk from the dataset. This change is real; you cannot get trunk back unless you cancel all your changes. (But remember, no changes are permanent until you save them on disk. You work with a copy of the data in memory, not with the data file itself.)

8. *Choose **Restore** to restore the backup copy of your data.*
   Click **Restore**. The dataset is now exactly as it was when you began. A backup copy of your dataset is automatically made when you enter the editor. (This can be changed; see [D] **edit** in the *Stata Data Management Reference Manual*.)

9. ***Preserve** updates the backup copy.*
   Sort again by mpg. Click on one of the cells that has mpg equal to 14. Click **Delete...**, select the third choice, "Delete all 6 obs. where mpg==14", and click **OK**.

   Click **Preserve**. Now make some other changes to the dataset. Click **Restore**. The changes that you made after the **Preserve** are reversed. There is no way to cancel the changes you made before you pressed **Preserve** (except to reload the original data file).

10. *When you exit the editor, a dialog will ask you to confirm your changes.*
    If you **Cancel** your changes, the editor reloads the backup copy.

11. *You will find output in the Results window documenting the changes you made.*
    What you see are Stata commands that are equivalent to what you did in the editor. The dash in front of the command indicates that the change was done in the editor.

```
. use "C:\Program Files\Stata9\auto.dta"
(1978 Automobile Data)

. edit
- preserve
- sort price
- sort weight
- drop trunk
- restore
- sort mpg
- drop if mpg == 14
- preserve
- replace weight = 10 in 5
- replace mpg = 10 in 10
- replace rep78 = . in 25
- drop in 3
- restore
```

edit with in and if

Here is one example of using `edit` with selected variables and restricted observations. Chapter 11 contains many examples of using a command with a variable list and `in` and `if`. Load `auto.dta`.

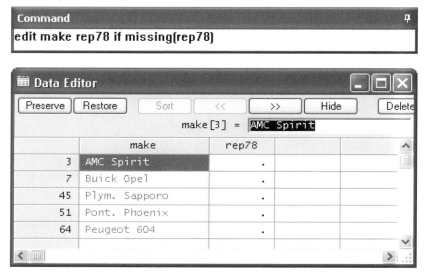

Notes

1. You cannot use the **Sort** button if you entered the editor with an `in` or `if` restriction.

2. Stata has multiple missing-value indicators: '.' is Stata's default or system missing-value indicator, and `.a`, `.b`, ..., `.z` are Stata's extended missing values. Some people use extended missing values to indicate why a certain value is unknown. Other people have no use for extended missing values and just use '.'.

   Stata views missing values as being larger than all numeric nonmissing values, and $. < .a <$ $.b < ... < .z$. `if rep78 >= .` is equivalent to `if missing(rep78)`. See [U] **12.2.1 Missing values** for full details.

Advice

1. *People who care about data integrity know that editors are dangerous—it is easy to accidentally make changes.* Never use `edit` when you just want to look at your data. Use `browse`.

2. *Protect yourself when you* `edit` *data by limiting the dataset's exposure.*
   If you need to change `rep78` only if it is missing, and need to see `make` to make the change, do not click the **Data Editor** button, but type

   ```
   edit make rep78 if missing(rep78)
   ```

   It is now impossible for you to change (damage) variables other than `make` and `rep78` and observations other than those with `rep78` equal to missing.

3. *All of this said, Stata's editor is safer than most because it records changes in the Results window.* Use this feature to log your output and make a permanent record of the changes. Then, you can verify that the changes you made are the changes you wanted to make. See chapter 16 for information on creating log files.

browse

When you want to look at your dataset but do not want to change it, you can enter the editor in browse mode.

1. *Load* auto.dta.

2. *Either click the* **Data Browser** *button, or type* browse *in the Command window.*

3. *In* browse *mode, it is impossible to alter your data.*

4. *You can select variables and restrict observations with* browse, *just like* with edit.

Command	ꟼ
browse make mpg price if foreign==1	

| Data Browser | | | | | | |

Preserve | Restore | Sort | << | >> | Hide | Delete...

make[53] = Audi 5000

	make	mpg	price	
53	Audi 5000	17	9,690	
54	Audi Fox	23	6,295	
55	BMW 320i	25	9,735	
56	Datsun 200	23	6,229	
57	Datsun 210	35	4,589	
58	Datsun 510	24	5,079	
59	Datsun 810	21	8,129	
60	Fiat Strada	21	4,296	
61	Honda Accord	25	5,799	

5. *You can use the* **Sort**, *shift* (<< *and* >>), *and* **Hide** *buttons in* browse *mode.*
   However, just like edit, you cannot sort if you entered browse with an in or if restriction.

6. *You can use* browse *to do much of what the* list *command does.*
   But, because you can scroll in the Data Editor, the editor is more convenient for viewing many variables and for scrolling up and down through the observations.

   The list command, which produces output in the Results window, is described in chapter 11. list is useful for producing listings for your log file (see chapter 16) and for viewing a few variables quickly in the Results window.

Right-clicking on the Data Editor window

Right-clicking on the editor window displays a menu from which you can select

---

- **Copy** to copy data to the Clipboard.
- **Paste** to paste data from the Clipboard.
- **Variable > Properties...** to display the **Variable Properties** dialog for the current variable.
- **Variable > Sort** to sort the current variable.
- **Variable > Hide** to hide the current variable.
- **Variable > Delete** to open the **Editor Delete Options** dialog.
- **Select Value from Value Label** '*name*' to display a menu of values for the value label *name*. Select an item from the menu to change the value for the current cell.
- **Assign Value Label to Variable** '*varname*' to display a menu of value labels. Select a value label from the menu to assign it to the current variable.
- **Define/Modify Value Labels...** to display the **Define Value Labels** dialog.
- **Hide All Value Labels...** to hide all value labels and display their numeric value instead.
- **Preferences...** to set the preferences for the editor.
- **Font** to change the font of the editor.

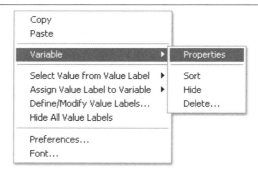

Exiting the Data Editor

1. *To exit the editor:*
   Click the editor's close box.

2. *Changes made in the editor are not saved until you tell Stata to save them.*
   The data that you have entered only exist in computer memory. They are not yet saved on disk.

3. *Save your dataset by selecting* **File > Save As...**.
   You will enter the filename thecars, and Stata then adds the extension .dta, so the file is saved as thecars.dta. See the next chapter for full details on saving your dataset.

4. *Note: You cannot save your dataset to disk until you exit the editor.*
   If you open the **File** menu while in the editor, **Save** and **Save As...** will be disabled.

Important implication: Changes are not permanent until you save them to disk. In Stata, you work with data in memory, not with data on a disk file. You can have Stata make a temporary backup of your data before making changes to it by clicking the **Preserve** button.

9 Inputting data from a file

insheet, infile, and infix

In this chapter, you will learn:

Stata can read text (ASCII) files.

There are 3 different commands for doing so: `insheet`
    `infile`
    `infix`

And one command for reading a SAS XPORT
Transport file: `fdause`

And one command for reading an ODBC source: `odbc`

And one command for reading an XML file: `xmluse`

If the text file was created by a spreadsheet
or a database program: Use `insheet`.

If the data in the text file are separated by
spaces and do not have string variables
(i.e., non-numeric characters), or if all
strings are just one word, or if all strings
are enclosed in quotes: Use `infile`.

Otherwise, you will have to specify the format
of the data in the text file: Use `infix` or `infile` (fixed format).

If you need to read datasets from the U.S. Food
and Drug Administration (FDA) in SAS
XPORT Transport format: Use `fdause`.

If you need to read data from a database that
supports ODBC: Use `odbc`.

If you need to read data from a Stata or Excel
XML file: Use `xmluse`.

Spreadsheets save text files with delimiters: `insheet` can read files delimited
with commas or tab characters.

`insheet` cannot read space-delimited files: Use `infile` as described on the next page.

`insheet` is easy to use: It will read your variable names from the file,
or it will make up variable names for you.

`insheet` is flexible: You can override the variable names it
would have chosen for you.

---

insheet, infile, and infix, continued

Before reading in data, you must clear memory.
    First save the current dataset (if you want it):   Select **File > Save**.
    Then clear memory by typing: `clear`

To read a text file created by a spreadsheet: `insheet using` *filename*
    To read `myfile.raw`:   `insheet using myfile`
    To read `myfile.xyz`:   `insheet using myfile.xyz`
    The extension `.raw` is assumed if you
    do not specify another extension.

Suppose that you have data for three numeric variables
    separated by spaces entered in a text file.
    To read them in with the names a, b, and c: `infile a b c using` *filename*
    To read from file `myfile.raw`:   `infile a b c using myfile`
    To read from file `myfile.xyz`:   `infile a b c using myfile.xyz`
    The extension `.raw` is assumed if you
    do not specify another extension.

Variables are given the datatype `float`
    unless you specify otherwise: a, b, c will be `float`s.

Stata has six datatypes for variables:

Real numbers, 8.5 digits of precision	`float`
Real numbers, 16.5 digits of precision	`double`

Integers between:
-127 and 100	`byte`
$-32,767$ and $32,740$	`int`
$-2,147,483,647$ and $2,147,483,620$	`long`

Strings (from 1 to 80 characters for Small Stata and
    Intercooled Stata; from 1 to 244 for Stata/SE)
1-character long strings	`str1`
2-character long strings	`str2`
3-character long strings	`str3`
$\vdots$	$\vdots$
244-character long strings	`str244`

insheet, infile, and infix, continued

The numeric variable types `byte` and `int` are used to reduce the memory that your dataset requires:	See [U] **12 Data** and [U] **6 Setting the size of memory**.
The numeric variable types `long` and `double` are used for special accuracy requirements:	See [U] **13.10 Precision and problems therein**.
You specify the datatype by putting it in front of the variable name. If you put nothing in front, the variable will be a `float`.	
To `infile` 3 variables:	`infile a b c using myfile`
If string `a` is 10 characters or less and `b` and `c` numeric:	`infile str10 a b c using myfile`
If `a` is numeric, `b` a string, `c` numeric:	`infile a str10 b c using myfile`
If `a` and `b` are numeric, `c` a string:	`infile a b str10 c using myfile`
If `a` and `b` are strings, `c` numeric:	`infile str10 a str10 b c using myfile`
or:	`infile str10 (a b) c using myfile`
Missing values:	
Missing numbers are indicated by: `.` is called the system missing value, and `.a`, `.b`, ..., and `.z` are called extended missing values.	`.`, `.a`, `.b`, ..., or `.z`
Missing strings are indicated by:	`""`
String values are in double quotes:	`"Harry Smith"`
If the string has no blanks, you can omit the quotes:	`Smith`
Stata can read formatted ASCII files. We give an example, but it is too complex to explain fully here:	See [D] **infile (fixed format)**, [D] **infix (fixed format)**, and [U] **21 Inputting data**.
Other programs can be purchased that convert other software formats to Stata format:	See [U] **21.4 Transfer programs**.

insheet

The `insheet` command was specially developed to read in text (ASCII) files that were created by spreadsheet programs. Many people have data that they enter into their spreadsheet programs. All spreadsheet programs have an option to save the dataset as a text (ASCII) file with the columns delimited with either tab characters or commas. Some spreadsheet programs also save the column titles (variable names in Stata) in the text file.

In order to read in this file, you only have to type `insheet using` *filename*, where *filename* is the name of the text file. The `insheet` command will determine whether there are variable names in the file, what the delimiter character is (tab or comma), and what type of data are in each column. If *filename* contains spaces, put double quotes around the filename.

If you have a text (ASCII) file that you created by saving data from a spreadsheet program, then try the `insheet` command to read that data into Stata.

Remember that by default the `insheet` command understands files that use the tab or comma as the column delimiter. If you have a file that uses spaces as the delimiter, use the `infile` command instead. See the `infile` section of this chapter for more information. If you have a file that uses another character as the delimiter, use `insheet`'s `delimiter()` option; see [D] **insheet** for more information.

Suppose that you have saved the file `sample.raw` from your favorite spreadsheet. It contains the following lines:

```
VW Rabbit,4697,25,1930,3.78
Olds 98,8814,21,4060,2.41
Chev. Monza,3667,,2750,2.73
,4099,22,2930,3.58
Datsun 510,5079,24,2280,3.54
Buick Regal,5189,20,3280,2.93
Datsun 810,8129,,2750,3.55
```

These correspond to the make, price, MPG, weight, and gear ratio of a few cars. Note that the variable names are not in the file (so `insheet` will assign its own names) and that the fields are separated by a comma. Use `insheet` to read the data:

```
. insheet using sample
(5 vars, 7 obs)

. list, separator(0)
```

	v1	v2	v3	v4	v5
1.	VW Rabbit	4697	25	1930	3.78
2.	Olds 98	8814	21	4060	2.41
3.	Chev. Monza	3667	.	2750	2.73
4.		4099	22	2930	3.58
5.	Datsun 510	5079	24	2280	3.54
6.	Buick Regal	5189	20	3280	2.93
7.	Datsun 810	8129	.	2750	3.55

insheet, continued

If you want to specify better variable names, you can include the desired names in the command:

```
. insheet make price mpg weight gear_ratio using sample
(5 vars, 7 obs)
. list, separator(0)
```

	make	price	mpg	weight	gear_r~o
1.	VW Rabbit	4697	25	1930	3.78
2.	Olds 98	8814	21	4060	2.41
3.	Chev. Monza	3667	.	2750	2.73
4.		4099	22	2930	3.58
5.	Datsun 510	5079	24	2280	3.54
6.	Buick Regal	5189	20	3280	2.93
7.	Datsun 810	8129	.	2750	3.55

Note

Stata listed gear_ratio as gear_r~o in the output from list. gear_r~o is a unique abbreviation for the variable gear_ratio. Stata displays the abbreviated variable name when variable names are longer than eight characters.

To prevent Stata from abbreviating gear_ratio, you could specify the abbreviate(10) option:

```
. list, separator(0) abbreviate(10)
```

	make	price	mpg	weight	gear_ratio
1.	VW Rabbit	4697	25	1930	3.78
2.	Olds 98	8814	21	4060	2.41
3.	Chev. Monza	3667	.	2750	2.73
4.		4099	22	2930	3.58
5.	Datsun 510	5079	24	2280	3.54
6.	Buick Regal	5189	20	3280	2.93
7.	Datsun 810	8129	.	2750	3.55

For more information on the ~ abbreviation and on list, see chapter 11.

infile

The file `afewcars.raw` contains the following lines:

```
"VW Rabbit"        4697         25       1930      3.78
"Olds 98"    8814    21   4060    2.41
"Chev. Monza"      3667          .        2750      2.73
" "                4099         22       2930      3.58
"Datsun 510"       5079         24       2280      3.54
"Buick Regal"
 5189
 20
 3280
 2.93
"Datsun 810"       8129        ***       2750      3.55
```

These correspond to the make, price, MPG, weight, and gear ratio of a few cars. Notice that the second line is not formatted the same as the first line; suppose that you do not know MPG on the third line (file contains '.'); you also do not know make on the fourth line (file contains ""); the sixth car's data are spread over 5 lines; and you do not know MPG on the last line (file contains '***').

```
. infile str18 make price mpg weight gear_ratio using afewcars
'***' cannot be read as a number for mpg[7]
(7 observations read)
. list, separator(0)

             make    price    mpg    weight    gear_r~o
  1.     VW Rabbit    4697     25      1930       3.78
  2.       Olds 98    8814     21      4060       2.41
  3.   Chev. Monza    3667      .      2750       2.73
  4.                  4099     22      2930       3.58
  5.    Datsun 510    5079     24      2280       3.54
  6.   Buick Regal    5189     20      3280       2.93
  7.    Datsun 810    8129      .      2750       3.55

. save afewcars.dta, replace
```

Notes

1. `using afewcars` means using `afewcars.raw`. The extension `.raw` is assumed if you do not specify otherwise. If the data had been stored in `afewcars.asc`, you would have had to specify `using afewcars.asc`.

2. As in the Data Editor, '.' indicates a numeric missing value.

3. As in the Data Editor, an empty string ("") means string missing value. However, with `infile` you must explicitly have a "" in your file so that it does not read the next number in the file as the value of this string variable.

4. Observations need not be on a single line. The location of line breaks does not matter.

5. Things that are not understood (such as ***) are mentioned and are stored as missing values.

infile with formatted data

File `cars2.raw` contains the following lines:

```
VW Rabbit        4697      25      1930    3.78
Olds 98          8814      21      4060    2.41
Chev. Monza      3667              2750    2.73
                 4099      22      2930    3.58
Datsun 510       5079      24      2280    3.54
Buick Regal      5189      20      3280    2.93
Datsun 810       8129              2750    3.55
```

These data are more difficult to read because (1) there are no double quotes around the strings in the first column and they include blanks and (2) there are blanks in the first and third columns when we do not know the value.

`infile`'s ordinary logic when reading five variables is

1st thing	in the file is	1st variable	in 1st observation
2nd thing		2nd variable	in 1st observation
3rd thing		3rd variable	in 1st observation
4th thing		4th variable	in 1st observation
5th thing		5th variable	in 1st observation
6th thing		1st variable	in 2nd observation
7th thing		2nd variable	in 2nd observation

$$\vdots \qquad\qquad \vdots \quad \vdots$$

Carry out this logic on the above:

VW	is make	in 1st observation	
Rabbit	is price	in 1st observation	(error, stores as missing value)
4697	is MPG	in 1st observation	(wrong)
25	is weight	in 1st observation	(also wrong)
1930	is gear ratio	in 1st observation	(wrong again)
3.78	is make	in 2nd observation	(stores "3.78" as a string!)
Olds	is price	in 2nd observation	(error, stores as missing value)
98	is MPG	in 2nd observation	(surprise!)

$$\vdots \qquad \vdots \qquad \vdots$$

This problem is referred to as "loss of synchronization". There is a solution.

infile with formatted data, continued

Using the Do-file Editor or a word processor, create `cars2.dct` containing the following lines:

```
dictionary using cars2.raw {
        _column(1)    str18 make      %11s
        _column(19)   price           %4f
        _column(31)   mpg             %2f
        _column(39)   weight          %4f
        _column(47)   gear_ratio      %4f
}
```

This is described in [D] **infile (fixed format)** in the *Stata Data Management Reference Manual*. To read the data,

```
. infile using cars2
dictionary using cars2.raw {
        _column(1)    str18 make      %11s
        _column(19)   price           %4f
        _column(31)   mpg             %2f
        _column(39)   weight          %4f
        _column(47)   gear_ratio      %4f
}
(7 observations read)
. list

          |        make    price    mpg    weight    gear_r~o
      1.  |    VW Rabbit     4697     25      1930        3.78
      2.  |      Olds 98     8814     21      4060        2.41
      3.  |  Chev. Monza     3667      .      2750        2.73
      4.  |                  4099     22      2930        3.58
      5.  |   Datsun 510     5079     24      2280        3.54

      6.  |   Buick Regal    5189     20      3280        2.93
      7.  |   Datsun 810     8129      .      2750        3.55
```

That is,

1. Create another file describing the contents of the dataset with another program—an editor or word processor. Name the file *myfile*.`dct`, and save it as a text (ASCII) file.

2. Instructions for creating the contents of this file can be found in [D] **infile (fixed format)**.

3. Once the dictionary exists, you type `infile using` *myfile*. The file extension .`dct` is assumed when `infile` is used without any variable names.

Importing files from other software

If you have a file created by another software package that you would like to read into Stata, you should be able to use options on that software package to output the file as a plain text (ASCII) file. Then, you can use `infile`, `insheet`, or `infix` to read in the data.

Or you may be able to copy and paste data from another spreadsheet into Stata's Data Editor. See [D] **edit** for full details.

Or you can use the `fdause` and `fdasave` commands for reading and writing SAS XPORT Transport files. See `fdause` and `fdasave` in [D] **fdasave** for full details.

Or you can use the `odbc` command. See [D] **odbc** for full details.

Or you can use the `xmluse` and `xmlsave` commands. See `xmluse` and `xmlsave` in [D] **xmlsave** for full details.

Or you can purchase a transfer program that will convert the other software's data file format to Stata's data-file format. See [U] **21.4 Transfer programs**.

Notes

10 Labeling data

describe and label

In this chapter, you will learn:

To obtain a description of the data currently in memory:	`describe`
stored on disk:	`describe using` *filename*
To set or reset the data label:	`label data "`*text*`"`
To set or reset a variable's label:	`label var` *varname* `"`*text*`"`
To label a variable's values:	
Create a value label:	`label define` *lblname* `#` `"`*text*`"` `#` `"`*text*`"` ...
Attach value label to variable:	`label values` *varname* *lblname*
You can attach the same value label to other variables, too.	
To unlabel	
the data:	`label data`
a variable:	`label var` *varname*
a variable's values:	`label values` *varname*
To delete a value label:	`label drop` *lblname*
To change a value label:	
Delete it:	`label drop` *lblname*
Redefine it:	`label define` *lblname* `#` `"`*text*`"` `#` `"`*text*`"` ...
To make changes permanent, resave the dataset:	select **File > Save**
or, alternatively, you can type:	`save` *filename*`,` `replace`

Note

We cover only the rudiments of value labels here. For instance, there is a way to modify value labels without deleting and redefining. See [U] **12.6.3 Value labels** for a complete treatment. You can also open the **Data > Labels** menu for a list of dialogs that can add, delete, define, or modify labels.

describe

In chapter 9, we saved a dataset called `afewcars.dta`:

```
. use afewcars

. list, separator(0)

              make    price   mpg   weight   gear_r~o
  1.      VW Rabbit    4697    25     1930      3.78
  2.         Olds 98   8814    21     4060      2.41
  3.    Chev. Monza    3667     .     2750      2.73
  4.                   4099    22     2930      3.58
  5.     Datsun 510    5079    24     2280      3.54
  6.    Buick Regal    5189    20     3280      2.93
  7.     Datsun 810    8129     .     2750      3.55
```

The description of this dataset is

```
. describe
Contains data from afewcars.dta
  obs:           7
  vars:          5                           9 May 2005 14:25
  size:        266 (99.9% of memory free)

              storage  display    value
variable name   type   format     label      variable label

make            str18   %18s
price           float   %9.0g
mpg             float   %9.0g
weight          float   %9.0g
gear_ratio      float   %9.0g

Sorted by:
```

1. The variable name is how we refer to the column of data.

2. The storage type is described in chapter 9.

3. The display format controls how the variable is displayed; see [U] **12.5 Formats: controlling how data are displayed**. By default, Stata sets it to something reasonable given the storage type.

Describing a file without loading it

It is not necessary to load a dataset before describing it:

```
. describe using afewcars
Contains data
  obs:             7                           9 May 2005 14:25
  vars:            5
  size:          266

              storage  display    value
variable name   type   format     label      variable label

make            str18  %18s
price           float  %9.0g
mpg             float  %9.0g
weight          float  %9.0g
gear_ratio      float  %9.0g

Sorted by:
```

That is,

1. When you type `describe` by itself, Stata describes the current contents of memory.

2. When you type `describe using` *filename*, Stata describes the contents of the specified Stata data file (i.e., a file named *filename*`.dta` created by Stata).

describe and labels

Datasets can contain labels for the dataset, variables, and values. An example is the dataset `auto.dta`, which is stored in the directory where you installed Stata. It is also stored online at the Stata Press web site, and Stata can even read it from that location! You will learn more about Stata's Internet capabilities in chapter 19.

```
. describe using http://www.stata-press.com/data/r9/auto
Contains data                          1978 Automobile Data
  obs:            74                   13 Apr 2005 17:45
 vars:            12
 size:         3,478

              storage  display   value
variable name   type   format    label      variable label

make           str18   %-18s                Make and Model
price          int     %8.0gc               Price
mpg            int     %8.0g                Mileage (mpg)
rep78          int     %8.0g                Repair Record 1978
headroom       float   %6.1f                Headroom (in.)
trunk          int     %8.0g                Trunk space (cu. ft.)
weight         int     %8.0gc               Weight (lbs.)
length         int     %8.0g                Length (in.)
turn           int     %8.0g                Turn Circle (ft.)
displacement   int     %8.0g                Displacement (cu. in.)
gear_ratio     float   %6.2f                Gear Ratio
foreign        byte    %8.0g     origin     Car type

Sorted by:  foreign
```

That is,

variable name		display format		variable label
↓		↙		↙
gear_ratio	float	%6.2f		Gear Ratio
foreign	byte	%8.0g	origin	Car type
	↑		↑	
	storage type		value label name	

1. Value labels allow numeric variables—such as `foreign`—to have words associated with numeric codes. The `describe` tells you that the numeric variable `foreign` has value label `origin` associated with it. Although not revealed by `describe`, the variable `foreign` takes on the values 0 and 1, and the value label `origin` associates 0 with Domestic and 1 with Foreign. When you `browse` the data (see chapter 8), `foreign` appears to contain the values "Domestic" and "Foreign". Note that it is not necessary for the value label to have a name different than the variable. You could just as easily have used a value label named `foreign`.

2. Variable labels are merely comments that help us remember the contents of variables. They are used by many Stata commands to enhance tabular and graphical output.

Labeling datasets and variables

You can easily label variables (in this case make and gear_ratio) and the dataset as a whole:

```
. use afewcars
. describe
Contains data from afewcars.dta
  obs:            7
  vars:           5                            9 May 2005 14:25
  size:         266 (99.9% of memory free)

              storage  display    value
variable name  type    format     label     variable label

make          str18    %18s
price         float    %9.0g
mpg           float    %9.0g
weight        float    %9.0g
gear_ratio    float    %9.0g

Sorted by:
. label var make "Make of car"
. label var gear_ratio "Gear ratio"
. label data "A few 1978 cars"
. describe
Contains data from afewcars.dta
  obs:            7                            A few 1978 cars
  vars:           5                            9 May 2005 14:25
  size:         266 (99.9% of memory free)

              storage  display    value
variable name  type    format     label     variable label

make          str18    %18s                  Make of car
price         float    %9.0g
mpg           float    %9.0g
weight        float    %9.0g
gear_ratio    float    %9.0g                  Gear ratio

Sorted by:
. save afewcars, replace
file afewcars.dta saved
```

That is,

1. Use label var to set or reset a variable's label. Put the label itself in double quotes:
   label var gear_ratio "Gear Ratio"

2. Use label data to set or reset the dataset's label.

3. You set or change the labels on the dataset in memory. To make the changes permanent, resave the dataset.

Labeling values of variables

Say that you have added the variable `foreign` to `afewcars.dta`. If you list the dataset, you will see

```
. list, separator(0)

             make    price    mpg    weight    gear_r~o    foreign
  1.     VW Rabbit     4697     25      1930        3.78          1
  2.        Olds 98    8814     21      4060        2.41          0
  3.    Chev. Monza    3667      .      2750        2.73          0
  4.                   4099     22      2930        3.58          0
  5.     Datsun 510    5079     24      2280        3.54          1
  6.    Buick Regal    5189     20      3280        2.93          0
  7.     Datsun 810    8129      .      2750        3.55          1
```

$foreign = 0$ denotes domestic cars, and 1 denotes foreign. To label the values "domestic" and "foreign", type

```
. label define origin 0 "domestic" 1 "foreign"
. label values foreign origin
. describe
Contains data from afewcars.dta
  obs:             7                        A few 1978 cars
  vars:            6                        9 May 2005 14:25
  size:          294  (99.8% of memory free)

                storage   display     value
variable name    type     format      label          variable label

make            str18    %18s                        Make of car
price           float    %9.0g
mpg             float    %9.0g
weight          float    %9.0g
gear_ratio      float    %9.0g                       Gear ratio
foreign         float    %9.0g       origin

Sorted by:
     Note:  dataset has changed since last saved
. save afewcars, replace
file afewcars.dta saved
```

That is,

1. Use `label define` to create a value label. The syntax is
    `label define` *labelname* `#` *"contents"* `#` *"contents"* ...

2. Use `label values` to associate the label with a variable. The syntax is

    `label values` *variablename labelname*

3. To make the change permanent, resave the dataset.

Labels and the GUI

You can open the **Data > Labels** menu for a list of dialogs that can add, delete, define, or modify labels. For example, to label a dataset, select **Data > Labels > Label dataset** to use the **Label dataset** dialog. To label a variable, select **Data > Labels > Label variable** to use the **Label variable** dialog.

For some commands, the dialogs are not any more efficient than typing the command by hand, but they may be helpful when you can't remember the syntax of the command. For example, instead of typing `label define` *labelname # "contents" # "contents"* ... to create a new value label, you could select **Data > Labels > Label values > Define or modify value labels** to open the **Define value labels** dialog. From this dialog,

1. Click the **Define...** button.
2. Type a label name, and click **OK**.
3. Type a numeric value in the **Value** edit field from the **Add value** dialog.
4. Type the value label in the **Text** edit field, and click **OK**.
5. Repeat steps 3 and 4.
6. Click **Cancel** when you are finished adding values.

You can always open the **Define value labels** dialog again to add or modify more value labels.

Notes

11 Listing data

list

In this chapter, you will learn:

---

list is similar to browse.
    To list data in the Results window, type: list

When a —more— appears in the Results window:
    (as happens when listing a long list)
    To see the next line: Press *Enter*.

    To see the next screen: Press any key, such as *Space Bar*.
    or: Click the **More** button.
    or: Click on the —more— at the
        bottom of the Results window.

To interrupt a Stata command at any time
    and return to the state before you
    issued the command: Click the **Break** button.
    or: Press q if you see a —more—

To list a single variable: list *varname*
    Example: list displacement
    You may abbreviate list: l displacement
    You may abbreviate *varname*:   list displ
    You may use a ~ abbreviation: list displa~t

To list any number of variables: list *varname(s)*
    Example: list make mpg
    You may abbreviate: l mak mpg

To list *varname$_i$* through *varname$_j$*:   list *varname$_i$*-*varname$_j$*
    Example: list make-mpg
    You may abbreviate: l mak-mpg

To list all variables starting with pop: list pop*

To list all variables starting with ma,
    a character, and ending in e: list ma?e

You may combine any of the above: list mak-mpg displa~t pop*

(Continued on next page)

list, continued

To list the third observation:	list in 3
the second from last:	list in -2
the last:	list in -1
or:	list in l (i.e., the letter ℓ)
To list observations 1 through 3:	list in 1/3
5 to 17:	list in 5/17
3 to third from last:	list in 3/-3
You may combine all the above:	list mpg pop* in 3/-3
To conditionally list observations:	list if *exp*
Example:	list if mpg>20
You may combine all the above:	list mpg wei pop* if mpg>20
	l mpg wei pop* if mpg>20 in 3/-3
Output from list can be logged (as can all output that appears in the Results window):	See chapter 16.
To specify a minimum abbreviation of 10 characters for variable names:	list, abbreviate(10)
To specify divider lines between columns in your listing:	list, divider
To specify horizontal separator lines be drawn after every 2 observations:	list, separator(2)
To prevent horizontal separator lines from being drawn:	list, separator(0)
To draw a horizontal separator line when variable foreign changes values:	list, sepby(foreign)

list with variable list

You can list a specified variable list. For example, you can list the make, mpg, and price variables. Or you might list variables beginning with the letter m or list all variables from price to weight.

```
. list

             make    price    mpg   weight   gear_r~o    foreign
  1.     VW Rabbit     4697     25     1930       3.78     foreign
  2.       Olds 98     8814     21     4060       2.41    domestic
  3.   Chev. Monza     3667      .     2750       2.73    domestic
  4.                   4099     22     2930       3.58    domestic
  5.     Datsun 510    5079     24     2280       3.54     foreign

  6.   Buick Regal     5189     20     3280       2.93    domestic
  7.   Datsun 810      8129      .     2750       3.55     foreign

. list make mpg price

             make    mpg    price
  1.     VW Rabbit     25     4697
  2.       Olds 98     21     8814
  3.   Chev. Monza      .     3667
  4.                   22     4099
  5.     Datsun 510    24     5079

  6.   Buick Regal     20     5189
  7.   Datsun 810       .     8129

. list m*

             make    mpg
  1.     VW Rabbit     25
  2.       Olds 98     21
  3.   Chev. Monza      .
  4.                   22
  5.     Datsun 510    24

  6.   Buick Regal     20
  7.   Datsun 810       .
```

list with variable list, continued

```
. list price-weight

        price    mpg    weight

  1.     4697     25     1930
  2.     8814     21     4060
  3.     3667      .     2750
  4.     4099     22     2930
  5.     5079     24     2280

  6.     5189     20     3280
  7.     8129      .     2750

. list ma?e

                     make

  1.          VW Rabbit
  2.            Olds 98
  3.        Chev. Monza
  4.
  5.         Datsun 510

  6.        Buick Regal
  7.        Datsun 810

. list gear_r~o

        gear_r~o

  1.         3.78
  2.         2.41
  3.         2.73
  4.         3.58
  5.         3.54

  6.         2.93
  7.         3.55
```

Notes

1. list without arguments lists the entire dataset.

2. You can list a subset of variables explicitly, as in list make mpg price, using shorthand: list m* means all variables starting with m; list price-weight means variables price through weight in dataset order; and list ma?e means all variables starting with ma, followed by any character, and ending in e.

3. You can list a variable using an abbreviation unique to that variable, as in list gear_r~o. If the abbreviation is not unique, Stata returns an error message.

4. You can abbreviate list as l (the letter ℓ).

list with in

```
. list

                 make     price    mpg    weight    gear_r~o     foreign

     1.       VW Rabbit     4697     25      1930        3.78      foreign
     2.         Olds 98     8814     21      4060        2.41     domestic
     3.     Chev. Monza     3667      .      2750        2.73     domestic
     4.                     4099     22      2930        3.58     domestic
     5.      Datsun 510     5079     24      2280        3.54      foreign

     6.     Buick Regal     5189     20      3280        2.93     domestic
     7.      Datsun 810     8129      .      2750        3.55      foreign

. list in 1

                 make     price    mpg    weight    gear_r~o     foreign

     1.     VW Rabbit     4697     25      1930        3.78      foreign

. list in -1

                 make     price    mpg    weight    gear_r~o     foreign

     7.     Datsun 810     8129      .      2750        3.55      foreign

. list in 2/4

                 make     price    mpg    weight    gear_r~o     foreign

     2.         Olds 98     8814     21      4060        2.41     domestic
     3.     Chev. Monza     3667      .      2750        2.73     domestic
     4.                     4099     22      2930        3.58     domestic

. list make mpg in -3/-2

                 make     mpg

     5.      Datsun 510     24
     6.     Buick Regal     20
```

Notes

1. in restricts the list to a range of observations.

2. Positive numbers count from the top of the dataset. Negative numbers count from the end of the dataset.

3. You may specify both a variable list and an observation range.

list with if

```
. list

              make    price    mpg   weight   gear_r~o    foreign

   1.      VW Rabbit     4697     25     1930       3.78    foreign
   2.        Olds 98     8814     21     4060       2.41   domestic
   3.     Chev. Monza     3667      .     2750       2.73   domestic
   4.                    4099     22     2930       3.58   domestic
   5.      Datsun 510     5079     24     2280       3.54    foreign

   6.     Buick Regal     5189     20     3280       2.93   domestic
   7.      Datsun 810     8129      .     2750       3.55    foreign

. list if mpg>22

              make    price    mpg   weight   gear_r~o    foreign

   1.      VW Rabbit     4697     25     1930       3.78    foreign
   3.     Chev. Monza     3667      .     2750       2.73   domestic
   5.      Datsun 510     5079     24     2280       3.54    foreign
   7.      Datsun 810     8129      .     2750       3.55    foreign

. list if mpg>22 & !missing(mpg)

              make    price    mpg   weight   gear_r~o    foreign

   1.      VW Rabbit     4697     25     1930       3.78    foreign
   5.      Datsun 510     5079     24     2280       3.54    foreign

. list make mpg if mpg>22 | (price>8000 & gear_ratio>3.5)

              make    mpg

   1.      VW Rabbit     25
   3.     Chev. Monza      .
   5.      Datsun 510     24
   7.      Datsun 810      .

. list make mpg if mpg>22 | (price>8000 & gear_ratio>3.5) in 1/4

              make    mpg

   1.      VW Rabbit     25
   3.     Chev. Monza      .
```

Notes

1. if expressions may be arbitrarily complicated. & means "and"; | means "or".

2. if may be combined with in (or in with if; the order does not matter).

list with if, continued

3. Stata supports different types of numeric missing values that can be used to specify different reasons a value is unknown. Numeric missing values are represented in Stata by "large positive values". Thus, mpg>22 evaluates to true when mpg equals missing value. You can exclude observations with missing values for mpg by adding '& mpg<.' to your if expression or by using the missing() function to test for missing values of mpg: '!missing(mpg)'. The '<' is the symbol for "less than", '.' is the missing-value symbol, and '!' means "not".

4. The logical operators are

<	less than
<=	less than or equal
==	equal
>	greater than
>=	greater than or equal
!=	not equal (~= can also be used)
&	and
|	or
!	not (logical negation; ~ can also be used)
()	parentheses specify order of evaluation

By default, & is evaluated before |; therefore, $a|b\&c$ means $a|(b\&c)$, which is true if a is true or if b and c are both true. To specify a or b being true, and c being true, too, type $(a|b)\&c$.

list with if, common mistakes

```
. list

              make    price    mpg    weight    gear_r~o    foreign

     1.     VW Rabbit   4697     25      1930        3.78      foreign
     2.       Olds 98   8814     21      4060        2.41     domestic
     3.    Chev. Monza   3667      .      2750        2.73     domestic
     4.                 4099     22      2930        3.58     domestic
     5.     Datsun 510   5079     24      2280        3.54      foreign

     6.    Buick Regal   5189     20      3280        2.93     domestic
     7.     Datsun 810   8129      .      2750        3.55      foreign

. list if mpg=21
=exp not allowed
r(101);

. list if mpg==21

              make    price    mpg    weight    gear_r~o    foreign

     2.       Olds 98   8814     21      4060        2.41     domestic

. list if mpg==21 if weight>4000
invalid syntax
r(198);

. list if mpg==21 and weight>4000
invalid 'and'
r(198);

. list if mpg==21 & weight>4000

              make    price    mpg    weight    gear_r~o    foreign

     2.       Olds 98   8814     21      4060        2.41     domestic
```

Notes

1. Tests of equality are specified with double equal signs, not single—if mpg==21, not if mpg=21. Single equal signs, you will learn, are used for assignment.

2. Joint tests are specified with &, not multiple ifs—if mpg==21 & weight>4000, not if mpg==21 if weight>4000.

3. Use & and |, not the words and and or.

list with if, common mistakes, continued

```
. list

              make     price    mpg   weight   gear_r~o     foreign

  1.      VW Rabbit     4697     25     1930       3.78      foreign
  2.        Olds 98     8814     21     4060       2.41     domestic
  3.    Chev. Monza     3667      .     2750       2.73     domestic
  4.                    4099     22     2930       3.58     domestic
  5.     Datsun 510     5079     24     2280       3.54      foreign

  6.    Buick Regal     5189     20     3280       2.93     domestic
  7.     Datsun 810     8129      .     2750       3.55      foreign

. list if make==Datsun 510
Datsun not found
r(111);

. list if make=="Datsun 510"

              make     price    mpg   weight   gear_r~o     foreign

  5.     Datsun 510     5079     24     2280       3.54      foreign
```

Note

Tests with strings are allowed, but if you specify the contents of the string, the contents are enclosed in double quotes: if make=="Datsun 510".

list with if, common mistakes, continued

```
. list if foreign==domestic
domestic not found
r(111);
. list if foreign==0
```

	make	price	mpg	weight	gear_r~o	foreign
2.	Olds 98	8814	21	4060	2.41	domestic
3.	Chev. Monza	3667	.	2750	2.73	domestic
4.		4099	22	2930	3.58	domestic
6.	Buick Regal	5189	20	3280	2.93	domestic

```
. list if foreign==0, nolabel
```

	make	price	mpg	weight	gear_r~o	foreign
2.	Olds 98	8814	21	4060	2.41	0
3.	Chev. Monza	3667	.	2750	2.73	0
4.		4099	22	2930	3.58	0
6.	Buick Regal	5189	20	3280	2.93	0

Notes

1. Watch out for value labels—they look like strings but are not. Variable foreign takes on values 0 and 1, and we have merely labeled 0 "domestic" and 1 "foreign" (see chapter 10).

2. To see the underlying numeric values of variables with labeled values, use the nolabel option on the list or edit commands (see [D] **list** and [D] **edit**). You can also determine the correspondence between labels and numeric values with the label list command (see [D] **label**).

Controlling the list output

Several options give you even more control over list output. You can use sepby() to separate observations by variable. abbreviate() specifies the minimum number of characters to abbreviate a variable in the output. divider draws a vertical line between the variables in the list.

```
. sort foreign
. list, sepby(foreign)
```

	make	price	mpg	weight	gear_r~o	foreign
1.		4099	22	2930	3.58	domestic
2.	Olds 98	8814	21	4060	2.41	domestic
3.	Buick Regal	5189	20	3280	2.93	domestic
4.	Chev. Monza	3667	.	2750	2.73	domestic
5.	Datsun 510	5079	24	2280	3.54	foreign
6.	Datsun 810	8129	.	2750	3.55	foreign
7.	VW Rabbit	4697	25	1930	3.78	foreign

```
. list make weight gear, abbreviate(10)
```

	make	weight	gear_ratio
1.		2930	3.58
2.	Olds 98	4060	2.41
3.	Buick Regal	3280	2.93
4.	Chev. Monza	2750	2.73
5.	Datsun 510	2280	3.54
6.	Datsun 810	2750	3.55
7.	VW Rabbit	1930	3.78

```
. list, divider
```

	make	price	mpg	weight	gear_r~o	foreign
1.		4099	22	2930	3.58	domestic
2.	Olds 98	8814	21	4060	2.41	domestic
3.	Buick Regal	5189	20	3280	2.93	domestic
4.	Chev. Monza	3667	.	2750	2.73	domestic
5.	Datsun 510	5079	24	2280	3.54	foreign
6.	Datsun 810	8129	.	2750	3.55	foreign
7.	VW Rabbit	4697	25	1930	3.78	foreign

Controlling the list output, continued

separator draws a horizontal line at specified intervals.

```
. list, separator(3)
```

	make	price	mpg	weight	gear_r~o	foreign
1.		4099	22	2930	3.58	domestic
2.	Olds 98	8814	21	4060	2.41	domestic
3.	Buick Regal	5189	20	3280	2.93	domestic
4.	Chev. Monza	3667	.	2750	2.73	domestic
5.	Datsun 510	5079	24	2280	3.54	foreign
6.	Datsun 810	8129	.	2750	3.55	foreign
7.	VW Rabbit	4697	25	1930	3.78	foreign

More

When you see a —more— at the bottom of the Results window, it means that there is more information to be displayed. This happens, for example, when you are listing a large number of observations.

```
. list make mpg

        make              mpg

  1.  Linc. Continental    12
  2.  Linc. Mark V          12
  3.  Linc. Versailles      14
  4.  Merc. XR-7            14
  5.  Cad. Deville          14

  6.  Peugeot 604           14
  7.  Cad. Eldorado         14
  8.  Merc. Cougar          14
  9.  Buick Electra         15
 10.  Merc. Marquis         15

 11.  Olds Toronado         16
 12.  Buick Riviera         16
 13.  Dodge Magnum          16
 14.  Chev. Impala          16
 15.  Volvo 260             17

 16.  Audi 5000             17
 17.  AMC Pacer             17
 18.  Dodge St. Regis       17
 19.  Pont. Firebird        18
 20.  Olds Delta 88         18

—more—
```

If you want to see the next screen of text, press any key, such as the *Space Bar*, click the **More** button, or click on the blue —more— at the bottom of the Results window. To see just the next line of text, press *Enter*.

Break

If you want to interrupt a Stata command, click the **Break** button. If you see a —more— at the bottom of the Results window and wish to interrupt it, click the **Break** button, or press q.

```
. list make mpg

     ┌─────────────────────────┐
     │ make              mpg   │
     ├─────────────────────────┤
  1. │ Linc. Continental  12   │
  2. │ Linc. Mark V       12   │
  3. │ Linc. Versailles   14   │
  4. │ Merc. XR-7         14   │
  5. │ Cad. Deville       14   │
     │                         │
  6. │ Peugeot 604        14   │
  7. │ Cad. Eldorado      14   │
  8. │ Merc. Cougar       14   │
  9. │ Buick Electra      15   │
 10. │ Merc. Marquis      15   │
     │                         │
 11. │ Olds Toronado      16   │
 12. │ Buick Riviera      16   │
 13. │ Dodge Magnum       16   │
 14. │ Chev. Impala       16   │
 15. │ Volvo 260          17   │
     │                         │
 16. │ Audi 5000          17   │
 17. │ AMC Pacer          17   │
 18. │ Dodge St. Regis    17   │
 19. │ Pont. Firebird     18   │
 20. │ Olds Delta 88      18   │
     └─────────────────────────┘
—Break—
r(1);
```

Note

It is always safe to click **Break**. After you click **Break**, the state of the system is the same as if you had never issued the command.

12 Creating new variables

generate and replace

In this chapter, you will learn:

To create a new variable that is an algebraic expression of other variables:	generate *newvar* = *exp*
generate may be abbreviated:	g *newvar* = *exp*
To change the contents of an existing variable:	replace *oldvar* = *exp*
replace may not be abbreviated.	
exp is an algebraic expression that is a combination of existing variables, operators, and functions.	

Operators:

	Arithmetic		Logical		Relational (numeric and string)
+	addition	!	not	>	greater than
−	subtraction	|	or	<	less than
*	multiplication	&	and	>=	> or equal
/	division			<=	< or equal
^	power			==	equal
				!=	not equal
+	string concatenation				

Stata has many mathematical, statistical, string, date, time-series, and programming functions. See [D] **functions** for a complete list and full details.

generate

generate creates a new variable that is an algebraic expression of other variables.

```
. list price mpg

        price    mpg

  1.     4697     25
  2.     8814     21
  3.     3667      .
  4.     4099     22
  5.     5079     24

  6.     5189     20
  7.     8129      .

. gen logpr = ln(price)
. gen ratio = price/mpg
(2 missing values generated)
. gen silly = ((price+100)/ln(mpg-3))^2
(2 missing values generated)
. list price mpg logpr ratio silly

        price    mpg      logpr      ratio      silly

  1.     4697     25   8.454679     187.88    2408405
  2.     8814     21   9.084097   419.7143    9511256
  3.     3667      .   8.207129          .          .
  4.     4099     22   8.318499   186.3182    2033699
  5.     5079     24   8.532869    211.625    2893700

  6.     5189     20   8.554296     259.45    3484886
  7.     8129      .   9.003193          .          .
```

Notes

1. The form of the generate command is generate *newvar* = *exp*, where *newvar* is a new variable name (it cannot be the name of a variable that already exists) and *exp* is any valid expression.

2. The generate command may be abbreviated as g, ge, gen, etc.

3. An expression is a combination of existing variables, operators, and functions. Expressions can be made as complicated as you want.

4. Calculation on a missing value yields a missing value, and so does division by zero, etc.

5. If missing values are generated, the number of missing values in *newvar* is always reported. The lack of a mention means no missing values resulted.

replace

You use `generate` to create new variables, and you use `replace` to change the contents of existing variables. Stata requires this so that you do not accidentally modify your data.

The `replace` command cannot be abbreviated. Stata generally requires you to spell out completely any command that can alter your existing data.

```
. list make weight price foreign, nolabel

                    make    weight    price    foreign

     1.        VW Rabbit      1930     4697          1
     2.          Olds 98      4060     8814          0
     3.      Chev. Monza      2750     3667          0
     4.                       2930     4099          0
     5.        Datsun 510     2280     5079          1

     6.       Buick Regal     3280     5189          0
     7.        Datsun 810     2750     8129          1

. gen weight = weight/1000
weight already defined
r(110);

. replace weight = weight/1000
(7 real changes made)
```

replace, continued

Say that you want to create a new variable predprice, which will be the predicted price of the cars in the following year. You estimate that domestic cars will increase in price by 5% and foreign cars by 10%. First use generate to compute the predicted domestic car prices. Then use replace to change the missing values for the foreign cars to their proper values.

```
. gen predprice = 1.05*price if foreign == 0
(3 missing values generated)
. replace predprice = 1.1*price if foreign == 1
(3 real changes made)
. list make weight price predprice foreign, nolabel
```

	make	weight	price	predpr~e	foreign
1.	VW Rabbit	1.93	4697	5166.7	1
2.	Olds 98	4.06	8814	9254.7	0
3.	Chev. Monza	2.75	3667	3850.35	0
4.		2.93	4099	4303.95	0
5.	Datsun 510	2.28	5079	5586.9	1
6.	Buick Regal	3.28	5189	5448.45	0
7.	Datsun 810	2.75	8129	8941.9	1

generate with string variables

Stata is smart. When you generate a variable and the expression evaluates to a string, Stata creates a string variable with a storage type as long as necessary, and no longer than that. `origin` is a `str1`.

```
. list make foreign, nolabel

              make    foreign

  1.      VW Rabbit          1
  2.        Olds 98          0
  3.    Chev. Monza          0
  4.                         0
  5.     Datsun 510          1

  6.    Buick Regal          0
  7.    Datsun 810          1

. gen origin = "D" if foreign == 0
(3 missing values generated)
. replace origin = "F" if foreign == 1
(3 real changes made)
. list make foreign origin

              make     foreign    origin

  1.      VW Rabbit     foreign         F
  2.        Olds 98    domestic         D
  3.    Chev. Monza    domestic         D
  4.                   domestic         D
  5.     Datsun 510     foreign         F

  6.    Buick Regal    domestic         D
  7.    Datsun 810     foreign         F

. describe origin

                 storage   display    value
variable name     type     format     label        variable label
───────────────────────────────────────────────────────────────────
origin            str1      %9s
```

generate with string variables, continued

```
. gen makeorigin = make + " " + origin

. gen word2 = substr(make, strpos(make," ")+1, .)
(1 missing value generated)

. list make origin makeorigin word2
```

	make	origin	makeorigin	word2
1.	VW Rabbit	F	VW Rabbit F	Rabbit
2.	Olds 98	D	Olds 98 D	98
3.	Chev. Monza	D	Chev. Monza D	Monza
4.		D	D	
5.	Datsun 510	F	Datsun 510 F	510
6.	Buick Regal	D	Buick Regal D	Regal
7.	Datsun 810	F	Datsun 810 F	810

Notes

1. The operator '+' when applied to string variables will concatenate the strings (i.e., join them together). The expression `"this"` + `"that"` results in the string `"thisthat"`. When the variable `makeorigin` was generated, a space (`" "`) was added between the two strings.

2. `strpos(`s_1`,`s_2`)` produces an integer equal to the first position in s_1 in which s_2 is found, or 0 if it is not found.

   `substr(`s`,`c_0`,`c_1`)` produces a string equal to the columns c_0 to $c_0 + c_1$ of s. If $c_1 = .$, it gives a string from c_0 to end of string.

   Therefore, `substr(`s`,strpos(`s`," ")+1,.)` produces s with its first word removed.

13 Deleting variables and observations

clear, drop, and keep

In this chapter, you will learn:

To drop all the data in memory:	`drop _all`
To drop all the data in memory and clear other memory areas used by graphs, dialogs, matrices, etc.:	`clear`
To drop a single variable:	`drop` *varname*
Example:	`drop weight`
Example:	`drop gear_r~o`
To drop any number of variables:	`drop` *varname*
Example:	`drop make mpg`
Example:	`drop ma?e`
To drop *varname*$_i$ through *varname*$_j$:	`drop` *varname*$_i$`-`*varname*$_j$
Example:	`drop make-mpg`
To drop all variables starting with pop:	`drop pop*`
You may combine all the above:	`drop make-mpg weight pop*`
To drop the first observation:	`drop in 1`
the second:	`drop in 2`
the third:	`drop in 3`
the second from last:	`drop in -2`
the last:	`drop in -1`
or:	`drop in l` (i.e., the letter ℓ)
To drop observations 1 through 3:	`drop in 1/3`
5 to 17:	`drop in 5/17`
3 to third from last:	`drop in 3/-3`
To conditionally drop observations:	`drop if` *exp*
Example:	`drop if mpg>20`
You may combine `if` and `in`:	`drop if mpg>20 in 3/-3`
`keep` works like `drop`, except you specify the variables or observations to keep:	`keep if mpg>20 in 3/-3`
To make changes permanent, resave the dataset:	Select **File > Save**.
Alternatively, you can type:	`save` *filename*`, replace`

drop _all and clear

Both `drop _all` and `clear` eliminate the dataset from memory:

```
. drop _all
. list
```

or

```
. clear
. list
```

They differ in that

1. `drop _all` drops just the dataset from memory.

2. `clear` drops the dataset, and it drops all value-label definitions, scalars and matrices, constraints and equations, programs, previous estimation results, dialogs, graphs, Mata objects, and class system objects. `clear`, in effect, resets Stata. The first time you use `clear` while you have a graph or dialog open, you may be surprised when that graph or dialog closes; this is necessary so that Stata can free all memory that is being used.

drop

You can use drop by itself and with if and in expressions.

```
. use afewcars
(A few 1978 cars)
. list
```

	make	price	mpg	weight	gear_r~o	foreign
1.	VW Rabbit	4697	25	1930	3.78	foreign
2.	Olds 98	8814	21	4060	2.41	domestic
3.	Chev. Monza	3667	.	2750	2.73	domestic
4.		4099	22	2930	3.58	domestic
5.	Datsun 510	5079	24	2280	3.54	foreign
6.	Buick Regal	5189	20	3280	2.93	domestic
7.	Datsun 810	8129	.	2750	3.55	foreign

```
. drop in 1/3
(3 observations deleted)
. list
```

	make	price	mpg	weight	gear_r~o	foreign
1.		4099	22	2930	3.58	domestic
2.	Datsun 510	5079	24	2280	3.54	foreign
3.	Buick Regal	5189	20	3280	2.93	domestic
4.	Datsun 810	8129	.	2750	3.55	foreign

```
. drop if mpg>21
(3 observations deleted)
. list
```

	make	price	mpg	weight	gear_r~o	foreign
1.	Buick Regal	5189	20	3280	2.93	domestic

```
. drop gear_ratio
. list
```

	make	price	mpg	weight	foreign
1.	Buick Regal	5189	20	3280	domestic

drop, continued

You can also use drop with wildcard characters.

```
. drop m*
. list

          price    weight     foreign

   1.      5189      3280     domestic

. drop _all
. list
```

Note

To make changes permanent, you must resave the dataset. Since we do not resave in the above example, afewcars.dta remains unchanged.

keep

keep tells Stata to drop all variables except those specified explicitly or through the use of an `if` or `in` expression.

```
. use afewcars, clear
(A few 1978 cars)
. list
```

	make	price	mpg	weight	gear_r~o	foreign
1.	VW Rabbit	4697	25	1930	3.78	foreign
2.	Olds 98	8814	21	4060	2.41	domestic
3.	Chev. Monza	3667	.	2750	2.73	domestic
4.		4099	22	2930	3.58	domestic
5.	Datsun 510	5079	24	2280	3.54	foreign
6.	Buick Regal	5189	20	3280	2.93	domestic
7.	Datsun 810	8129	.	2750	3.55	foreign

```
. keep in 4/7
(3 observations deleted)
. list
```

	make	price	mpg	weight	gear_r~o	foreign
1.		4099	22	2930	3.58	domestic
2.	Datsun 510	5079	24	2280	3.54	foreign
3.	Buick Regal	5189	20	3280	2.93	domestic
4.	Datsun 810	8129	.	2750	3.55	foreign

```
. keep if mpg<=21
(3 observations deleted)
. list
```

	make	price	mpg	weight	gear_r~o	foreign
1.	Buick Regal	5189	20	3280	2.93	domestic

Note

To make changes permanent, you must resave the dataset. Since we do not resave in the above example, `afewcars.dta` remains unchanged.

Notes

14 Using the Do-file Editor

The Do-file Editor

In this chapter, you will learn:

Stata has a Do-file Editor.
   You can edit do-files and other text files with it.

To enter the Do-file Editor:
   - Click the **Do-file Editor** button
   - Or type `doedit` and press *Enter* in the Command window

The Do-file Editor lets you submit several commands to Stata at once.
   - Type the commands into the editor
   - Then click the **Do** button

The Do-file Editor has standard features found in other text editors.
   Cut, **Copy**, **Paste**, **Undo**, **Open**, **Save**, and **Print** are just a few examples.

The Do-file Editor has more advanced features to aid you in writing do-files.
   You will learn about them in this chapter.

You may have multiple Do-file Editors open.

The Do-file Editor toolbar

The Do-file Editor has twelve buttons. If you ever forget what a button
does, hold the mouse pointer over a button for a moment, and a box will
appear with a description of that button.

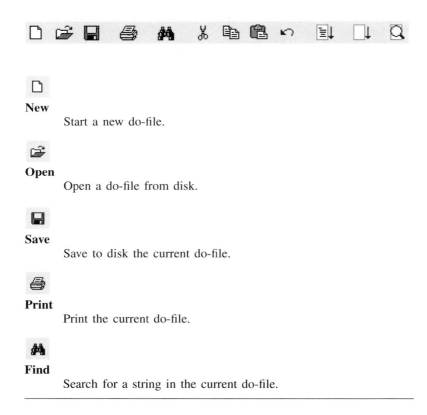

New

Start a new do-file.

Open

Open a do-file from disk.

Save

Save to disk the current do-file.

Print

Print the current do-file.

Find

Search for a string in the current do-file.

The Do-file Editor toolbar, continued

Cut

Copy the selected text to the Clipboard and cut it from the current do-file.

Copy

Copy the selected text to the Clipboard.

Paste

Paste text from the Clipboard into the current do-file.

Undo

Undo the last change.

Do

Execute the commands in the current do-file.

Run

Execute the commands in the current do-file without showing any output.

Preview

Preview the current file in the Viewer. The current file must be saved first.

Using the Do-file Editor

Suppose that you are about to do an analysis on fuel usage for 1978 automobiles. You know that you will be issuing many commands to Stata during your analysis and that you want to be able to reproduce your work later without having to type each of the commands again.

You may place commands in a text file; Stata can then read the file and execute each command in sequence. This file is known as a Stata "do-file"; see [U] **16 Do-files**.

To analyze fuel usage of 1978 automobiles, you plan to create a new variable giving gallons per mile. You want to see how that variable changes in relation to vehicle weight for both domestic and imported cars. You decide that performing a regression would be a good first step. You want to issue the following commands to Stata:

```
use http://www.stata-press.com/data/r9/auto, clear
generate gpm = 1/mpg
label var gpm "Gallons per mile"
sort foreign
regress gpm weight foreign
```

Using the Do-file Editor, continued

Click the **Do-file Editor** button to open the Do-file Editor. Then type the commands that you wish to submit to Stata. Note that the variable name on the fourth line is misspelled—go ahead and intentionally misspell it.

When you are through typing, click the **Do** button:

Using the Do-file Editor, continued

When you click the **Do** button, Stata executes the commands in sequence, and the results appear in the Results window:

```
. do "C:\temp\STD000001.tmp"
. use http://www.stata-press.com/data/r9/auto, clear
(1978 Automobile Data)
. generate gpm = 1/mpg
. label var gpm "Gallons per mile"
. sort foreing
variable foreing not found
r(111);
end of do-file
r(111);
```

The do "C:\..." is how Stata executes the commands that you typed in the Do-file Editor. Stata saves the commands to a temporary file and issues the do command to execute them.

Everything worked as planned until Stata saw the misspelled variable. The first three commands were executed, but an error was produced on the fourth. Stata does not know of a variable named foreing. Go back to the Do-file Editor, and correct the mistake.

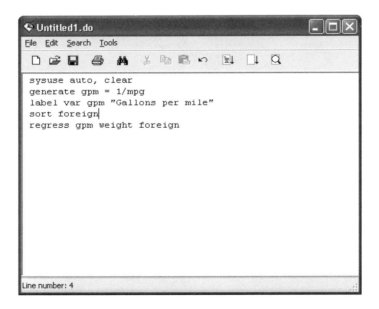

Using the Do-file Editor, continued

After you correct the spelling of `foreign` and click the **Do** button, Stata will execute each command in the editor, from the beginning.

```
. do "C:\temp\STD000001.tmp"
. use http://www.stata-press.com/data/r9/auto, clear
(1978 Automobile Data)
. generate gpm = 1/mpg
. label var gpm "Gallons per mile"
. sort foreign
. regress gpm weight foreign

      Source |       SS       df       MS              Number of obs =      74
-------------+------------------------------           F(  2,    71) =  113.97
       Model |  .009117618     2   .004558809           Prob > F      =  0.0000
    Residual |   .00284001    71      .00004           R-squared     =  0.7625
-------------+------------------------------           Adj R-squared =  0.7558
       Total |  .011957628    73   .000163803          Root MSE      =  .00632

------------------------------------------------------------------------------
         gpm |      Coef.   Std. Err.      t    P>|t|     [95% Conf. Interval]
-------------+----------------------------------------------------------------
      weight |   .0000163   1.18e-06    13.74   0.000     .0000139    .0000186
     foreign |   .0062205   .0019974     3.11   0.003     .0022379    .0102032
       _cons |  -.0007348   .0040199    -0.18   0.855    -.0087504    .0072807
------------------------------------------------------------------------------

.
end of do-file
```

You might want to select **File > Save As...** from the Do-file Editor to save this do-file. Later, you could select **File > Open...** to open it and then add more commands as you move forward with your analysis. By saving the commands of your analysis in a do-file as you go, you do not have to worry about retyping them with each new Stata session.

The File menu

The **File** menu of the Do-file Editor includes standard features found in most text editors. You may choose to start a **New** file, **Open...** an existing file, **Save** the current file, or save the current file under a new name with **Save As...**. You may also **Print...** the current file. There are also buttons on the Do-file Editor's toolbar that correspond to these features.

There is one other useful feature under the **File** menu: you may select **Insert File...** to insert the contents of another file at the current cursor position in the Do-file Editor.

Editing tools

The **Edit** menu of the Do-file Editor includes the standard **Cut**, **Copy**, and **Paste** capabilities, along with a single-level **Undo**. There are also buttons on the Do-file Editor's toolbar for easy access to these capabilities. There are several other **Edit** features that you may find useful.

You may delete or select the current line.

You may also shift the current line or selection right or left one tab stop.

The **Change Case** menu item allows you to change the case of the character to the right of the cursor or to change an entire selection. When you **Change Case** of an entire selection, the editor will compare the number of lowercase characters with the number of uppercase characters. The case with the greater number of characters will be changed. For example, if you have aBcDe selected, you will have ABCDE after selecting **Change Case**. The editor sees that there are fewer uppercase (2) than lowercase (3) characters, so it makes the entire selection uppercase. If you select **Change Case** again, the entire selection will change to lowercase. There are fewer lowercase (0) than uppercase (5) characters, so all are switched to lowercase.

If there is no selection when you select **Change Case**, the case of the character immediately to the right of the cursor is switched. The cursor is also moved one character to the right. For example, if the cursor is to the left of 'a' in aBcDe when you select **Change Case**, you will see ABcDe and the cursor will be just to the left of 'B'. If you select **Change Case** again, you will see AbcDe, and the cursor will be just to the left of 'c'.

Preferences

When you right-click on the Do-file Editor window and select **Preferences...**, you may customize the way the editor behaves.

You may set the number of spaces each *Tab* character represents. You may also set whether the Do-file Editor automatically indents the current line to the same level of indentation as the previous line.

You may change the font, and you may set whether or not the Do-file Editor automatically saves the current file when you click the **Do** button or the **Run** button. Normally, the file on which you are working is not saved to disk unless you explicitly select **File > Save** or **File > Save As...**. For example, if you open C:\data\myfile.do and make some changes, those changes will not be saved until you save them from the **File** menu. The changes are not saved, even when you execute myfile.do by clicking the **Do** button or the **Run** button. This prevents your original file from being overwritten until you tell the Do-file Editor to overwrite it. If you check **Auto-save on Do/Run** in the **Preferences** dialog, any changes that you have made to myfile.do will be saved when you click either the **Do** button or the **Run** button.

Searching

Stata's Do-file Editor includes standard **Find...** and **Replace...** capabilities in the **Search** menu. You should already be familiar with these capabilities from other text editors and word processors. You may access the **Find** capability from a button on the toolbar as well.

The **Search** menu also includes some choices that are useful when you are writing a long do-file or ado-file. You may select **Search > Go to Line ...** and jump straight to any line in your file.

Matching and balancing of parentheses ( ), braces { }, and brackets [ ] are available from the **Search** menu. When you select **Search > Match**, the Do-file Editor looks at the character immediately to the right of the cursor. If it is one of the characters that the editor can match, the editor will find the matching character and place the cursor immediately in front of it. If there is no match, you will hear a beep and the cursor will not move.

When you select **Search > Balance**, the Do-file Editor looks to the left and right of the current cursor position or selection and creates a selection including the narrowest level of matching characters. If you select **Balance** again, the editor will expand the selection to include the next level of matching characters. If there is no match, you will hear a beep and the cursor will not move.

Balance is easier to explain with an example. Type (now (is the) time) in the Do-file Editor. Place the cursor between the words is and the. Select **Search > Balance**. The Do-file Editor will select (is the). If you select **Balance** again, the Do-file Editor will select (now (is the) time).

The Tools menu

You have already learned about the **Do** button. Clicking it is equivalent to choosing **Tools > Do** in the Do-file Editor.

Next to the **Do** button is the **Run** button. Clicking it is equivalent to choosing **Tools > Run**. **Run** executes all the commands in the Do-file Editor just like **Do** does, but **Run** suppresses output. It is unlikely that you will ever need to use **Run**; see [U] **16.6.2 Suppressing output** for more details.

Do and **Run** are equivalent to Stata's do and run commands; see [U] **16 Do-files** for a complete discussion.

There are two other useful choices on the **Tools** menu. If you wish to execute a subset of the lines in your do-file, highlight those lines, and select **Tools > Do Selection**.

If you wish to execute all commands from the current line through the end of the file, select **Tools > Do to Bottom**.

You can also preview files in the Viewer by selecting **Tools > Preview file in Viewer** or by clicking the **Preview** button

Previewing files in the Viewer

You can use the Viewer to preview files open in the Do-file Editor by selecting **Tools > Preview File in Viewer** or by clicking the **Preview** button on the Do-file Editor's toolbar. This is useful for viewing SMCL files, such as Stata help files. Before a file may be previewed, it must be saved first (Stata will prompt you to save the file).

Saving interactive commands from Stata as a do-file

While working interactively with Stata, you may decide that you would like to rerun the last several commands that you typed interactively. You can save the contents of the Review window as a do-file and open that file in the Do-file Editor. You can copy a command from a dialog (rather than submit it) and paste it into the Do-file Editor. See chapter 4 for details. Also see [R] **log** for information on the cmdlog command, which allows you to log all commands that you type in Stata to a do-file.

15 Graphs

Working with graphs

In this chapter, you will learn

You can have multiple graph windows open.

The **Graph** button brings the topmost Graph window to the front of all other windows.

To print a graph:
- The Graph window must be open
- Right-click on the Graph window, and select **Print...**
- Or click on the Graph window, and then click the **Print** button on the toolbar
- Or select **File > Print >** *"graph title"*

To save a graph to disk:
- The Graph window must be open
- Right-click on the Graph window, and select **Save Graph...**
- Enter the *filename*
- Click **OK**

If you do not specify an extension, .gph will be added, and the file will be named *filename*.gph.

You can export a graph as a Windows Metafile (WMF), PNG, TIFF, PostScript, or EPS file by right-clicking on the Graph window, selecting **Save Graph...**, and selecting the format from the **Save as type:** list.

To load a graph from disk:
- Select **File > Open Graph...**
- Select a graph filename
- Click **OK**

- Or double-click on the .gph file from the Desktop
- Or type graph use *filename*
- Or type graph use (do not press *Enter*),
  choose **Filename...** under the **File** menu, select the file,
  and then press *Enter*

There is much to learn about Stata graphs; read the *Graphics Reference Manual* for full details.

The Graph button

The **Graph** button is located on the main window's toolbar. The button has two parts, an icon and an arrow. Clicking on the icon brings the topmost Graph window to the front of all other windows. Clicking on the arrow displays a menu of open graphs. Selecting a graph from the menu brings the graph to the front of all other windows.

If you ever decide to close the Graph window, you can only reopen it by reissuing a Stata command that draws a new graph.

Saving and printing graphs

You can save a graph once it is displayed by right-clicking on its window and selecting **Save Graph...**. You can print a graph by right-clicking on its window and selecting **Print...**. You can also use the **File** menu to save or print a graph. We recommend that you always right-click on a graph to save or print it to ensure that the correct one is selected.

For more information about printing graphs, see the *Stata Graphics Reference Manual*.

Right-clicking on the Graph window

Right-clicking on the Graph window displays a menu from which you can select

- **Save Graph...** to save the graph to disk.
- **Copy** to copy the graph to the Clipboard.
- **Preferences...** to edit the preferences for graphs.
- **Print...** to print the contents of the Graph window.

16 Logs: Printing and saving output

Using logs in Stata

In this chapter, you will learn:

A log is a recording of your Stata session.

Logs are stored as Stata formatted (SMCL) files or text (ASCII) files.
- You can view them in the Viewer
- You can print them
- You can translate formatted (SMCL) logs into text (ASCII) logs
- You can load ASCII logs into a text editor or word processor as you would any other text file

To start a log:
- Click the **Begin Log** button, and then enter a *filename*

Or:
- Select **File > Log > Begin...**
- Select the log type
- Enter a *filename*

Or:
- Type log using *filename*

The log file will be named
- *filename*.smcl for SMCL logs
- *filename*.log for ASCII logs

If you specify a file that already exists, you will be asked to choose to
- View the existing file
- Append the new log to the file
- Overwrite the file with the new log

The log file is closed automatically when you exit Stata.

To close the log file sooner:
- Click **Close/Suspend Log**, and choose **Close log file**
- Or type: log close

To temporarily suspend output from being written to the log:
- Click **Close/Suspend Log**, and choose **Suspend log file**
- Or type: log off

To resume the suspended log:
- Click **Close/Resume Log** and choose **Resume**
- Or type: log on

(Continued on next page)

139

Using logs in Stata, continued

To view a snapshot of the current log in the Viewer:
- Select **File > Log > View...**
- The filename of the current log is filled in automatically

To view an updated snapshot of the current log in the Viewer:
- Click the **Refresh** button at the top of the Viewer

To print the current log:
- View a snapshot of the current log in the Viewer
- Right-click on the Viewer window and select **Print...**

Comments can be added to your log as you work:
- In the Command window or in a do-file, type a '*' at the beginning of the line:
  ```
  * this is a comment
  * report last regression
  ```

Alternatively, you can use the log command to create a log file.

To start a new log file:
- Type: log using *filename*
- Type: log using *filename*, smcl
- Type: log using *filename*, text

To append to *filename*.smcl or *filename*.log:
- Type: log using *filename*, append

To overwrite *filename*.smcl or *filename*.log:
- Type: log using *filename*, replace

To translate a SMCL log to a text log:
- Select **File > Log > Translate...**
- Enter the SMCL log file in the **Input File** field
- Enter the new file in the **Output File** field
- Click **Translate**

To create a log that contains only the command lines that you type:
- Type: cmdlog using *filename*
- This log will be saved as *filename*.txt
- You can run the do-file to reproduce the output by typing do *filename*.txt

Logging output

All the output that appears in the Results window can be captured in a log file. The log file can be saved as a Stata formatted (SMCL) file or as a text (plain ASCII) file. The formatted file can be printed from Stata, retaining the **bolds**, <u>underlines</u>, and *italics* that you see in the Results window. The text file can be printed from Stata or loaded into a text editor or word processor.

1. *To start a log file, click the* **Begin Log** *button and enter a name for the file.*
   This will open a standard file dialog allowing you to specify a directory and filename to save your log. If you do not specify a file extension, the extension .smcl will be added to the filename.

2. *If you specify a file that already exists:*
   You will be asked if you want to append the new log to the file, or else overwrite the file with the new log.

3. *A Stata-formatted (SMCL) log file contains the output that you saw during your Stata session. A text (ASCII) log file will contain the ASCII translation of the output that you saw during your Stata session.*

```
. log using "C:\data\base.smcl"
        log:  C:\data\base.smcl
   log type:  smcl
 opened on:   12 May 2005, 17:04:05
. use http://www.stata-press.com/data/r9/auto
(1978 Automobile Data)
. sort foreigh
variable foreigh not found
r(111);
. sort foreign
. by foreign: summarize price mpg

-> foreign = Domestic
    Variable |      Obs        Mean    Std. Dev.       Min        Max
-------------+--------------------------------------------------------
       price |       52    6072.423    3097.104       3291      15906
         mpg |       52    19.82692    4.743297         12         34

-> foreign = Foreign
    Variable |      Obs        Mean    Std. Dev.       Min        Max
-------------+--------------------------------------------------------
       price |       22    6384.682    2621.915       3748      12990
         mpg |       22    24.77273    6.611187         14         41
. * include the above means in my report
. log close
        log:  C:\data\base.smcl
   log type:  smcl
 closed on:   12 May 2005, 17:04:13
```

4. *You can scroll through the Viewer to view previous output.*

Logging output, continued

5. *To display the log in the Viewer:* Select **File > Log > View....** Note: The log file image displayed in the Viewer is a snapshot taken at the time the Viewer was opened. To update the log image, click the **Refresh** button at the top of the Viewer.

6. *You can copy from the Viewer to the Clipboard and paste into the Command window:* Highlight the text that you want, right-click on the Viewer, and select **Copy**. To paste into the Command window, right-click on the Command window, and select **Paste**. Note: If text from the Clipboard contains more than one line, only the first line is pasted into the Command window.

7. *You can also copy from the Viewer to the Clipboard and paste into the Do-file Editor:* Highlight the text that you want, right-click on the Viewer, and select **Copy**. To paste into the Do-file Editor, right-click on the Do-file Editor, and select **Paste**.

8. *When you copy from a SMCL file in the Viewer, the format of the copied material will be in plain text (ASCII) format.*

9. *You can add comments to your log during your Stata session.*
   If you type a '*' at the beginning of a command line, the line is treated as a comment. Comments can be very helpful when you later read your log file and try to understand what you did (and why you did it).

```
. regress mpg weight wtsq gear_ratio displacement

  .
  . (output omitted)
  .
. regress mpg weight wtsq displacement

  .
  . (output omitted)
  .
. regress mpg weight wtsq

  .
  . (output omitted)
  .
. * the above is the regression I want to report
```

10. *The log file is closed automatically when you exit Stata.*

11. *Or you can close it whenever you want:*
    Click the **Close/Suspend Log** button, and choose **Close log file**.

12. *For more information about logs, see* [U] **15 Printing and preserving output** *and* [R] **log**. *For more information about the Viewer, see* chapter 5.

Printing logs

1. *To print a log file from the Viewer during a Stata session:*
    Right-click on the Viewer window, and select **Print...**, or select the Viewer window from **File > Print >** *"viewer title"*. After you click **OK**, a **Print** dialog will appear. After you click **Print**, an **Output Settings** dialog will appear.

    You can fill in none, any, or all of the items "Header", "Name", and "Project". You can check or uncheck options to **Print Line #s**, **Print Header**, and **Print Logo**. These items are saved and will appear again in the **Output Settings** dialog (in this and in future Stata sessions).

    You can set the margins and color scheme that the printer will use by clicking on **Prefs...** in the **Output Settings** dialog to open the **Printer Preferences** dialog. Monochrome is for black-and-white printing, Color is for default color printing, and Custom 1 and Custom 2 are for customized color printing.

    You can set the font that the printer will use from the **Printer Preferences** dialog. The font dialog will only list the fixed-width "typewriter" fonts (e.g., Courier) available for your printer. Stata, by default, will choose a font size that it thinks is appropriate for your printer.

2. *To print a closed log file:*
    To print a closed log file, you must open it first. Select **File > View...**, enter the filename, and click **OK**. Then, print the log file according to the above instructions. You could also use the `translate` command to generate a Postscript file. See [R] **translate** for more information.

3. *If your log file is a text file (*`.log`* instead of* `.smcl`*), you can load it into a text editor (such as Notepad), the Do-file Editor in Stata, or your favorite word processor.*

    From within your editor or word processor, you can edit the log file—add headings, comments, etc.—format it, and print it.

    Note about word processors:
    Your word processor will print out the file in its default font. The log file will not be very readable when printed in a proportionally spaced font (e.g., Times Roman or Helvetica). It will look much better printed in a fixed-width "typewriter" font (e.g., Courier).

    You may wish to associate the `.log` extension with a text editor (such as Notepad, Write, or WordPad) in Windows. You can then edit and print the logs from those Windows applications if you like.

    In Windows, you associate an extension with an application by clicking once on a file with the appropriate extension and then choosing **Options** from the **View** menu. In the **Options** dialog box, click on the **File Types** tab. See the documentation that came with your operating system for more information.

    Once you have done this, anytime you double-click on a file with that extension, the application associated with the extension will be opened and the file loaded into the application.

4. *If your log file is a Stata formatted file (*`.smcl`* instead of* `.log`*), you will need to open it in the Viewer to be able to see the output the way that it is displayed in the Results window.*

Rerunning commands as do-files

1. *To create a log file that contains only the command lines that you enter in a Stata session, type*

   `cmdlog using myfile`

2. *No output of any kind—no error messages, etc.—appears in a log file created with* `cmdlog`. Here's what the log shown on page 141 would look like if it had been created with the `cmdlog` option:

   ```
   use http://www.stata-press.com/data/r9/auto
   sort foreigh
   sort foreign
   by foreign: summarize price mpg
   *include the above means in my report
   ```

3. *To close the log, type*

   `cmdlog close`

4. *You can also save the contents of the Review window as a do-file.* The Review window stores the last 100 commands that you typed. Simply right-click on the Review window, and select **Save Review Contents...** from the menu. You may find this a more convenient way to create a text file containing only the commands that you typed during your session.

5. *You can edit a file created with* `cmdlog` *and rerun the commands in a batch mode from within Stata.* Such files are called do-files, since the batch-mode execution is done with Stata's do command. Here's how to create and run a do-file:

   a. Edit the file if necessary. In the previous example, we would want to delete the mistyped command `sort foreigh`. You can open the log file in the Do-file Editor in Stata or in your favorite text editor. When the cmdlog file is closed, the command `cmdlog close` is entered but is not included in the do-file.

   b. If you have any long command lines—ones that wrap around to two or more lines—you will have to make a few simple changes to your file. See [U] **16.1.3 Long lines in do-files** for details. If none of your command lines extends past one line, you need not be concerned about this.

   c. If you are using a word processor to edit the file, be sure to save the file as a text (plain ASCII) file. If you used the `cmdlog` option to make the file, the default extension will be `.do`.

   d. To rerun the commands, enter Stata, and type do *filename*. Alternatively, you could select **File > Do...** and then select the file that you want to execute in the dialog.

      Note: When you run a do-file, the output may continuously scroll without stopping for a —more—, so if you want to see it, you should either begin a new log or insert some more commands in your do-file. See [U] **7 —more— conditions**.

   e. See [U] **16 Do-files** and [U] **15 Printing and preserving output** for more information.

17 Setting font and window preferences

Changing and saving fonts and positions of your windows

In this chapter, you will learn:

You can change fonts in the following windows:

Results	(fixed-width fonts only)
Graph	(TrueType fonts only)
Viewer	(fixed-width fonts only)
Command	(any font)
Review	(any font)
Variables	(any font)
Data Editor	(fixed-width fonts only)
Do-file Editor	(fixed-width TrueType fonts only)

(Fixed-width fonts are "typewriter" fonts like Courier.)

To change the font for a window:
- Right-click on the window, select **Font...** to open a standard Windows font dialog, and select the font and size
- In the Do-file Editor, right-click on the window, select **Preferences...**, and select the font and size

The next time Stata is launched, it will use these preferences.

Stata supports multiple sets of saved preferences.

(*Continued on next page*)

Closing and opening windows

You can close all windows but the Results and Command windows. If you want to open a closed window, open the **Window** menu, and select the desired window.

Changing color schemes

To change the color scheme of the Results window:
- • Right-click on the Results window, and select **Preferences...**
- • Select the desired scheme from the **Color scheme** list

Or:
- • Create your own by selecting Custom 1 from the **Color scheme** list
- • Click the various output level buttons, and select the desired color
- • Check the **bold** and **underline** boxes if desired

To change the color scheme of the Viewer:
- • Right-click on the Viewer, and select **Preferences...**
- • Select the desired scheme from the **Color scheme** list

Or:
- • Create your own by selecting Custom 1 from the **Color scheme** list
- • Click the various output level buttons, and select the desired color
- • Check the **bold** and **underline** boxes if desired

Managing multiple sets of preferences

Stata's preferences are automatically saved after Stata has exited and are reloaded when Stata is launched. However, sometimes you may wish to rearrange Stata's windows and then revert the windows back to your preferred arrangement. You can do this by saving your preferences to a named preference set and loading them later. Any changes you make to Stata's preferences after loading a preference set do not affect the set; the set remains untouched unless you specifically overwrite it.

To load and save your preferences, open the **Prefs > Manage Preferences** menu, and
- • select a preference set from the **Load Preferences** menu to load it.
- • select **Load Preferences > Factory Settings** to restore the preferences to their factory settings. Other default windowing settings can be found in the **Load Preferences** menu.
- • select **Save Preferences > New Preferences Set...** to save the current preferences to a set. Enter a name for the set, and click **OK**.
- • select an existing set from the **Save Preferences** menu to overwrite it with the current preferences.
- • select a preference set from the **Delete Preferences** menu to delete it. Click **OK** to verify that you wish to delete the set.

18 How to learn more about Stata

Where to go from here

In this chapter, you will learn:

You should now know enough to begin using Stata.

Play with Stata: It's the best way to learn.

If you make a mistake and cannot figure out exactly what's wrong, look at the command's help file: Select **Help > Stata Command...**, and enter *commandname*.

Look at the syntax diagram and examples in the help file, and compare them with what you typed.

If you want to find the Stata command that produces a particular statistic, try: Select **Help > Search...**, choose **Search documentation and FAQs**, and enter *topic*.

Also, try the index in [I] .

Read the: *Stata User's Guide*

The *User's Guide* is designed to be read cover to cover, and contains much useful information about Stata.

The Reference manuals are not designed to be read cover to cover, but to be sampled when needed. For example, if you plan to do logistic regressions, read about the `logistic` command: [R] **logistic**

The datasets used in the examples in the Reference manuals are posted at: *http://www.stata-press.com/data/*

You can try performing the many examples in these manuals using the example datasets.

(*Continued on next page*)

Where to go from here, continued

This chapter contains some advice and
    recommended reading lists for sampling the
    User's Guide and the Reference manuals.

Look at the Stata web site;
    it has much useful information, including
    answers to frequently asked questions (FAQs):   *http://www.stata.com/support/faqs/*

Many useful links to Stata resources are found at: *http://www.stata.com/links/resources.html*

Do you want to learn more?
    Take a Stata NetCourse™:   NetCourse 101 is an excellent choice
        to learn about Stata. See
        http://www.stata.com/netcourse/
        for course information and schedules.

Suggested reading from the User's Guide and Reference manuals

The *User's Guide* is designed to be read from cover to cover in a linear fashion. The Reference manuals are designed as references to be sampled when necessary.

Ideally, after reading this *Getting Started* manual, you should read the *User's Guide* from cover to cover, but you probably want to become at least somewhat proficient in Stata right away. Here is a suggested reading list of sections from the *User's Guide* and the Reference manuals to help you on your way to becoming a Stata expert.

This list covers fundamental features and also points you to some less-obvious features that you might otherwise overlook.

Basic elements of Stata
[U]	Chapter 11	Language syntax
[U]	Chapter 12	Data
[U]	Chapter 13	Functions and expressions

Memory
[U]	Chapter 6	Setting the size of memory
[D]	compress	Compress data in memory

Data input
[U]	Chapter 6	Setting the size of memory
[U]	Chapter 21	Inputting data
[D]	edit	Edit and list data using Data Editor
[D]	infile	Quick reference for reading data into Stata
[D]	insheet	Read text (ASCII) data created by a spreadsheet
[D]	append	Append datasets
[D]	merge	Merge datasets

Graphics

Stata Graphics Reference Manual

Useful features that you might overlook
[U]	Chapter 28	Using the Internet to keep up to date
[U]	Chapter 16	Do-files
[U]	Chapter 19	Immediate commands
[U]	Chapter 23	Dealing with strings
[U]	Chapter 24	Dealing with dates
[U]	Chapter 25	Dealing with categorical variables
[U]	Chapter 13.5	Accessing coefficients and standard errors
[U]	Chapter 13.6	Accessing results from Stata commands

Suggested reading, continued

Estimation commands

[U] Chapter 26	Overview of Stata estimation commands
[U] Chapter 20	Estimation and postestimation commands
[U] Chapter 13.5	Accessing coefficients and standard errors
[R] estimates	Estimation results

Basic statistics

[R] anova	Analysis of variance and covariance
[R] ci	Confidence intervals for means, proportions, and counts
[R] correlate	Correlations (covariances) of variables or estimators
[D] egen	Extensions to generate
[R] regress	Linear regression
[R] predict	Obtain predictions, residuals, etc., after estimation
[R] regress postestimation	Postestimation tools for regress
[R] test	Test linear hypotheses after estimation
[R] summarize	Summary statistics
[R] table	Tables of summary statistics
[R] tabulate oneway	One-way tables of frequencies
[R] tabulate twoway	Two-way tables of frequencies
[R] ttest	Mean comparison tests

Matrices

[U] Chapter 14	Matrix expressions
[U] Chapter 18.5	Scalars and matrices

Mata Reference Manual

Programming

[U] Chapter 16	Do-files
[U] Chapter 17	Ado-files
[U] Chapter 18	Programming Stata
[R] ml	Maximum-likelihood estimation

Stata Programming Reference Manual

System values

[R] set	Overview of system parameters
[P] creturn	Return c-class values

Internet resources

The Stata web site (*http://www.stata.com*) is a good place to get more information about Stata. Half of the web site is dedicated to user support. You will find answers to frequently asked questions (FAQs), ways to interact with other users, official Stata updates, and other useful information. You can also subscribe to Statalist, a list server devoted to Stata and statistics discussion.

In addition, you will find information on Stata NetCourses™, which are interactive courses offered over the Internet and vary in length from a few weeks to eight weeks. Visit the Stata web site for more information.

At the web site is a Bookstore that contains books that we feel may be of interest to Stata users. Each book has a brief description written by a member of our technical staff explaining why we think this book may be of interest.

If you have access to the web, we suggest that you take a quick look at the Stata web site now. You can register your copy of Stata online and request a free subscription to the *Stata News*.

Visit *http://www.stata-press.com* for information on books, manuals, and journals published by Stata Press. The datasets used in examples in the Stata manuals are available from the Stata Press web site.

Also visit *http://www.stata-journal.com* to read about the *Stata Journal*, a quarterly publication containing articles about statistics, data analysis, teaching methods, and effective use of Stata's language.

See chapter 19 for details on accessing official Stata updates and free additions to Stata on the Stata web site.

Notes

19 Using the Internet

Internet functionality in Stata

In this chapter, you will learn:

To use Stata to contact Stata's web site for the latest StataCorp news:
- Select **Help > News**
- Or type: `news`

To query Stata's web site to see if you have the latest official updates:
- Select **Help > Official Updates**
- Click on `http://www.stata.com`
- Or type: `update query`
- Or let Stata automatically check for updates

To use Stata to visit web sites that have new commands you can download:
- Select **Help > SJ and User-written Programs**
- Click on `Search...`
- Or type: `search` *keyword*`, net`

To install a *Stata Journal* insert:
- Select **Help > SJ and User-written Programs**
- Click on `Stata Journal`

And then to install insert st0012 from SJ2-2
- Click on `sj2-2`
- Click on `st0012`
- Click on `click here to install`

To install an STB insert:
- Select **Help > SJ and User-written Programs**
- Click on `STB`

And then to install insert sg89 from STB-44
- Click on `stb44`
- Click on `sg89`
- Click on `click here to install`

(Continued on next page)

Internet functionality in Stata, continued

To load a dataset that a colleague has placed on a web page:
- Type: use http://www.stata.com/man/example.dta

To load a dataset from a previously set web site:
- Type: webuse cancer.dta
- The default web site is http://www.stata-press.com/data/r9/

To copy a dataset over the web from a colleague:
- Type: copy http://www.stata.com/man/example.dta mycopy.dta

To look at a text file on a web page:
- Type: type http://www.stata.com/man/readme.txt

To copy a text file over the web:
- Type: copy http://www.stata.com/man/readme.txt readme.txt

Official Stata

By official Stata, we mean the pieces of Stata that are provided and supported by StataCorp. The other and equally important pieces are the user-written additions published in the SJ, distributed over Statalist, or distributed in other ways.

Stata can fetch both official updates and user-written programs from the Internet. In the first case, you choose **Help > Official Updates**. In the second case, you choose **Help > SJ and User-written Programs**. (There are commands for doing this, too; see [U] **28 Using the Internet to keep up to date**.)

Let's start with the official updates. There are two parts to official Stata: the Stata executable and Stata's ado-files.

StataCorp releases updates to official Stata often. These updates are to add new features and, sometimes, to fix bugs. Typically the updates are to the ado-files because, in fact, most of Stata is written in Stata's ado language. Occasionally, we update the executable as well.

When you choose **Official Updates**, Stata tells you the dates of its two official pieces:

```
update

Stata executable
    folder:                 C:\Program Files\Stata9\
    name of file:           wsestata.exe
    currently installed:    12 May 2005
Ado-file updates
    folder:                 C:\Program Files\Stata9\ado\updates\
    names of files:         (various)
    currently installed:    12 May 2005
Recommendation
    compare these dates with what is available from
        http://www.stata.com
        other location of your choosing
        cdrom drive
```

If you know of a location that has official updates other than the Stata web site, click on `other location of your choosing`, and fill in the site or directory name that you wish.

If you are connected to the Internet, you can click on `http://www.stata.com` to compare the dates of your files with the latest updates available on the Stata web site. If you have trouble connecting to the Internet from Stata, visit *http://www.stata.com/support/faqs/web/* for help.

Official Stata, continued

When you click on http://www.stata.com, Stata compares your two official pieces with the most up-to-date versions available:

```
update query

(contacting http://www.stata.com/)
Stata executable
    folder:                  C:\Program Files\Stata9\
    name of file:            wsestata.exe
    currently installed:     12 May 2005
    latest available:        12 May 2005
Ado-file updates
    folder:                  C:\Program Files\Stata9\ado\updates\
    names of file:           (various)
    currently installed:     12 May 2005
    latest available:        12 May 2005
Recommendation
    Do nothing; all files up to date.
```

In this case, all files are up to date.

You might be told that you need to update your ado-files, your executable, or both. Entering update all will handle all cases. The sections **Updating the official ado-files** and **Updating the executable** are for informational purposes only. You should never update one without also updating the other if both updates are available. An update is not complete until you follow all the instructions presented on the screen.

Automatic update checking

Stata can periodically check for updates for you. By default, Stata will check once every seven days for updates from Stata's web site. The seven-day interval is from the last time an update query was performed regardless of whether it was by Stata or by you. You can change the interval between checks.

Before Stata connects to the Internet to check for an update, it will ask you if you would like to check now, check the next time Stata is launched, or check after the next interval. You can disable the prompt and allow Stata to check without asking.

If an update is available, Stata will notify you. From there, you should follow the recommendations for updating Stata.

You can change the settings for automatic update checking by selecting **Prefs > General Preferences...** and choosing **Internet**.

Updating the official ado-files

Much of Stata is implemented as ado-files—text files containing programs in Stata's ado language. StataCorp periodically releases updates. The **Official Updates** system makes it easy to keep your official ado-files up to date.

After selecting **Help > Official Updates** and then clicking on `http://www.stata.com`, you might see

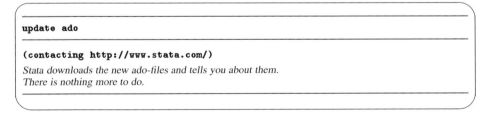

```
update query

(contacting http://www.stata.com/)
Information about other types of updates may appear . . .
Ado-file updates
    folder:                  C:\Program Files\Stata9\ado\updates\
    names of files:          (various)
    currently installed:     12 May 2005
    latest available:        18 Jun 2005
Recommendation
    update ado-files
```

Stata compares the dates of your ado-files with those available from StataCorp. In this case, Stata is recommending that you update your official ado-files.

If you do not have write permission for `C:\Program Files\Stata9`, you cannot install official updates in this way. You may still download the official updates, but you will need to use the command-line version of `update`; see [U] **28 Using the Internet to keep up to date** for instructions.

Click on `update ado-files`:

```
update ado

(contacting http://www.stata.com/)
Stata downloads the new ado-files and tells you about them.
There is nothing more to do.
```

For your information, Stata does not overwrite your original ado-files, which are installed in the directory `C:\Program Files\Stata9\ado\base`. Instead, Stata downloads updates to the directory `C:\Program Files\Stata9\ado\updates`. If there were ever a problem with the updates, you could simply remove the files in `C:\Program Files\Stata9\ado\updates`.

Updating the executable

The **Official Updates** system also checks your executable against the most current available:

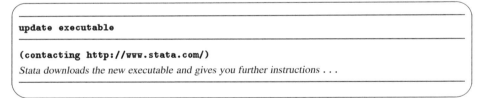

```
update query

(contacting http://www.stata.com/)
Stata executable
    folder:                C:\Program Files\Stata9\
    name of file:          wsestata.exe
    currently installed:   12 May 2005
    latest available:      03 Jun 2005
Information about other types of updates may appear . . .
Recommendation
    update executable
```

In this case, Stata recommends that you update your executable.

If you do not have write permission for C:\Program Files\Stata9, you cannot install official updates in this way. You may still download the official updates, but you will need to use the command-line version of update; see [U] **28 Using the Internet to keep up to date** for instructions.

Click on update executable:

```
update executable

(contacting http://www.stata.com/)
Stata downloads the new executable and gives you further instructions . . .
```

Stata copies the new executable to C:\Program Files\Stata9\wsestata.exe.bin; it does not replace your existing executable. That is for you to do next. Exit all instances of Stata that are running except the one in which you ran update. You must follow the instructions presented by Stata to complete the update.

Finding user-written programs by keyword

Stata has a built-in utility created specifically to search the Internet for user-written Stata programs. You can access it by selecting **Help > Search...**, choosing **Search net resources**, and entering a *keyword* in the field. Equivalently, you can select **Help > SJ and User-written Programs**, choose **Search on Internet...**, and enter a *keyword* in the field. The utility searches all user-written programs on the Internet, including the entire collection of *Stata Journal* and STB programs. The results are displayed in the Viewer, and you can click to go to any of the matches found.

For the syntax on how to use the equivalent search *keywords*, net command, see [R] **search**.

Downloading user-written programs

Downloading user-written programs is easy. Start by selecting **Help > SJ and User-written Programs**:

```
Installation and maintenance of SJ, STB, and user-written programs

    User-written programs -- SJ, STB, Statalist, and others -- are available
    from a variety of sources.  Use the links below to find, install, and
    uninstall them.

    New packages

        Search...               Search all sources.  Try this first!

        Stata Journal           Programs from Stata Journal articles
        STB                     Programs from Stata Technical Bulletin articles
        Statalist archive       Boston College Statalist program archive
        Other locations         Other locations with available programs

        Advanced
                enter site name...
                cdrom drive

    Previously installed packages
        List
        Search...

Also see

    Manual:   [U] 28 Using the Internet to keep up to date,
              [R] net
    Online:   help for net, search, sj, stb, update
```

What to do next will be obvious.

Downloading user-written programs, continued

For instance, `lfitx2` is a command from STB-44 sg87. Your Stata has no such command:

```
. lfitx2
unrecognized command:  lfitx2
r(199);
```

You might, however, discover that you want `lfitx2` because you used Stata's `search` command and it sounded interesting.

```
. search windmeijer goodness of fit
STB-44   sg87 . . . . Windmeijer's goodness-of-fit test for logistic regression
         (help lfitx2 if installed) . . . . . . . . . . . . . . . J. Weesie
         7/98    pp.22--27; STB Reprints Vol 8, pp.153--160
         alternative to lfit
```

To obtain sg87 from STB-44, click on `http://www.stata.com` after selecting **Help > SJ and User-written Programs**.

You should then click on `stb`, which will present you with a listing of all STBs available from StataCorp. Click on `stb44`:

```
http://www.stata.com/stb/stb44/
STB-44 July 1998

DIRECTORIES you could -net cd- to:
    . .                 Other STBs
PACKAGES you could -net describe-:
    dm59               Collapsing datasets to frequencies
    sbe19_1            Tests for publication bias in meta-analysis
    sbe24              metan -- an alternative meta-analysis command
    sg85               Moving summaries
    sg86               Continuation-ratio models for ordinal response data
    sg87               Windmeijer's goodness-of-fit test for logistic regression
    sg88               Estimating generalized ordered logit models
    sg89               Adjusted predictions and probabilities after estimation
    ssa12              Predicted survival curves for the Cox model
```

Downloading user-written programs, continued

You may click on any insert to obtain more information. Click on sg87:

```
package sg87 from http://www.stata.com/stb/stb44

TITLE
     STB-44 sg87.  Windmeijer's goodness-of-fit test for logistic regression.
DESCRIPTION/AUTHOR(S)
     STB insert by Jeroen Weesie, Utrecht University, Netherlands.
     Support: weesie@weesie.fsw.ruu.nl
     After installation, see help lfitx2.
INSTALLATION FILES                               (click here to install)
     sg87/lfitx2.ado
     sg87/lfitx2.hlp
ANCILLARY FILES                                  (click here to get)
     sg87/lbw.dta
```

Click on `click here to install` to download and install the `lfitx2.ado` and `lfitx2.hlp` files from sg87. (If you want the `lbw.dta` file associated with sg87, click on `click here to get`.)

You now have `lfitx2`. Select **Help > Stata command**, and enter `lfitx2`:

```
help for lfitx2                                  (STB-44: sg87)

Goodness-of-fit after logistic
------------------------------

          lfitx2 [, eps(#)]
The rest of the help file is displayed . . .
```

You can install lots of commands.

You can find out what you have installed by selecting **Help > SJ and User-written Programs** and clicking on `List` under `Previously installed packages`.

Try it after first downloading the STB insert:

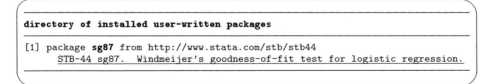

```
directory of installed user-written packages

[1] package sg87 from http://www.stata.com/stb/stb44
     STB-44 sg87.  Windmeijer's goodness-of-fit test for logistic regression.
```

Downloading user-written programs, continued

If you click on the one-line description of the insert, you will see the full description of the package that you installed:

```
package sg87 from http://www.stata.com/stb/stb44

TITLE                                    (click here to uninstall)
    STB-44 sg87.  Windmeijer's goodness-of-fit test for logistic regression.
DESCRIPTION/AUTHOR(S)
    STB insert by Jeroen Weesie, Utrecht University, Netherlands.
    Support: weesie@weesie.fsw.ruu.nl
    After installation, see help lfitx2.
INSTALLATION FILES
    l/lfitx2.ado
    l/lfitx2.hlp
INSTALLED ON
    13 May 2005
```

You can uninstall materials by clicking on `click here to uninstall` when you are looking at the package description.

Try it.

For information on downloading user-written programs using the `net` command, see [R] **net**.

A More on starting and exiting Stata

Contents

(*Continued on next page*)

A.1 Starting Stata

We assume that Stata is already installed. If you have not yet installed Stata, follow the installation instructions in chapter 1. To start Stata:

1. Open the Windows **Start** menu, and open the **Stata 9** program group.

2. Select **Stata/SE**, **Intercooled Stata**, or **Small Stata** as appropriate; see [U] **5 Flavors of Stata**.

You will see something like this:

Stata is now waiting for you to type something in the Command window. If Stata does not launch, see [GSW] **B.1 If Stata does not start**.

A.2 Verifying that Stata is correctly installed

The first time that you start Stata, you should verify that it is installed correctly. Type `verinst` in the Command window, and you should see something like

```
. verinst
You are running Stata/SE 9.0 for Windows.
Stata is correctly installed.
You can type exit to exit Stata.
```

If you see an error message complaining that the command was not found, then not all of Stata is installed. Go back to chapter 1 and reinstall Stata.

If you see any other error message, see [GSW] **B.2 verinst problems**.

Remember the `verinst` command. If you ever change your computer setup and are worried that you somehow damaged Stata in the process, you can type `verinst` and obtain the reassuring "Stata is correctly installed" message.

A.3   Exiting Stata

To exit Stata:

1. Click on the close box

    Stata works with a copy of the dataset in memory. If (1) there is a dataset in memory, and if (2) it has changed, and if (3) you click to close, a box will appear asking if it is okay to exit without saving the changes.

2. Or type `exit` in the Command window.

    If you instead type `exit` and there are changed data in memory, Stata will refuse and instead say, "no; data in memory would be lost". In this case, either you must save the dataset on disk (see [D] **save**), or you can type `exit, clear` if you do not want to save the changes.

All of this is designed to prevent you from accidentally losing your data.

As you will discover, the `clear` option is allowed with all potentially destructive commands, of which `exit` is just one example. The command to bring a dataset into memory, `use`, is another example. `use` is destructive since it loads the dataset into memory and, in the process, eliminates the dataset already there.

If you type a destructive command and the dataset in memory has been safely stored on disk, Stata performs your request. If your dataset has changed in some way since it was last saved, Stata responds with the message "no; data in memory would be lost". If you want to go ahead anyway, you can retype the command and add the `clear` option. Once you become familiar with Stata's editing keys, you will discover that it is not necessary to physically retype the line. You can press the *PrevLine* key (*PgUp*) to retrieve the last command you typed and append , `clear`.

Of course, you need not wait for Stata to complain—you can add the `clear` option the first time you issue the command—if you do not mind living dangerously.

A.4   The Windows Properties Sheet

When you double-click on a shortcut to start an application in Windows, you are actually executing instructions defined in the shortcut's Properties Sheet. To open the Properties Sheet for any shortcut, right-click on the shortcut, and select **Properties**.

Open the Properties Sheet for Stata's shortcut. Click on the **Shortcut** tab. You will see something that contains the following information:

Stata/SE

Target type:	Application
Target location:	Stata 9
Target:	`C:\Program Files\Stata9\wsestata.exe`
Start in:	`C:\data\`
Shortcut key:	`None`
Run:	`Normal window`

The field names may be slightly different, depending on the version of Windows you are running. The names and locations of files may vary from this. There are two things to pay attention to: The **Target** and **Start in**. **Target** is the actual command that is executed to invoke Stata. **Start in** is the directory to switch to before invoking the application. You can change these fields and then click **OK** to save the updated Properties Sheet.

A.5   Starting Stata from other folders

You can have Stata start in any directory you desire. Just change the **Start in** field of Stata's Properties Sheet.

Of course, once Stata is running, you can change directories whenever you wish; see [D] **cd**.

A.6   Specifying the amount of memory allocated

You can add a /m option in the **Target** field of Stata's Properties Sheet to set the initial amount of memory allocated to Stata:

    Target: C:\Program Files\Stata9\wsestata.exe /m20

Only Stata/SE and Intercooled users can do this; Small Stata has no /m, and the memory allocation is fixed.

The /m20 in the **Target** field tells Stata to allocate 20 megabytes of memory for data.

If you now want Stata to allocate 2 megabytes of memory, change the /m20 to /m2.

If you want 32 megabytes, change the field to /m32.

If you want 100 megabytes, change the field to /m100.

Note: if you make the amount too large, when Stata attempts to launch, you will see a message telling you that the operating system refused to provide that much memory. Stata will then attempt to launch using its default setting of 10 megabytes for Stata/SE or 1 megabyte for Intercooled Stata. If you see this message, you should decrease the amount of memory that you are asking the operating system to provide to Stata.

You can also change the amount of memory once Stata is running by using the set memory command, and you can make that setting permanent rather than modifying the /m option on the **Target**; see [GSW] **C. Setting the size of memory**. We recommend using the set memory method to set the memory allocation, as it is easier to use.

A.7   Executing commands every time Stata is started

Stata looks for the file `profile.do` when it is invoked and, if it finds it, executes the commands in it. Stata looks for `profile.do` first in the directory where Stata is installed, then in the current directory, then along your PATH, then in your home directory as defined by Windows' USERPROFILE environment variable, and finally along the adopath (see [P] **sysdir**); we recommend that you put `profile.do` in your working directory, `C:\data`.

Say that every time you started Stata, you wanted `matsize` set to 100 (see [R] **matsize**) . Create file `C:\data\profile.do` containing

```
set matsize 100
```

When you invoke Stata, this command will be executed:

```
(usual opening appears, but with the addition)
running C:\data\profile.do ...
. _
```

Note that you could also type `set matsize 100, permanently` in Stata. The `permanently` option tells Stata to remember the setting for future sessions and eliminates the need to put the `set matsize` command in `profile.do`.

`profile.do` is treated just as any other do-file once it is executed; results are literally the same as if you started Stata and then typed 'run `profile.do`'. The only special thing about `profile.do` is that Stata looks for it and runs it automatically.

See [U] **16 Do-files** for an explanation of do-files. They are nothing more than text (ASCII) files containing a sequence of commands for Stata to execute.

A.8   Making shortcuts

You can arrange to start Stata without going through the **Start** menu by creating a shortcut on the Desktop. The easiest way to do this is to copy the existing Stata shortcut to the Desktop. You can also create a shortcut directly from the Stata executable. Here are the details:

1. Open the `C:\Program Files\Stata9` folder or the folder where you installed Stata.

2. In the folder, find the executable for which you want a new shortcut. The filenames are

Stata/SE:	`wsestata.exe`
Intercooled Stata:	`wstata.exe`
Small Stata:	`wsmstata.exe`

Right-click and drag the appropriate executable onto the Desktop.

3. Release the mouse button, and select **Create Shortcut(s) Here** from the menu that appears.

You have now created a shortcut. If you want the shortcut in a folder rather than on the Desktop, you can drag it into whatever folder appeals to you.

You set the properties for this shortcut just as you would normally. Right-click on the shortcut, and select **Properties**. Edit the Properties Sheet as explained in [GSW] **A.4 The Windows Properties Sheet**. Some users create separate shortcuts for starting Stata with differing amounts of memory.

A.9 Executing Stata in background (batch) mode

You can run large jobs in Stata in batch mode. To do so, open a DOS window, change to your data directory, and type

```
C:\data> "C:\Program Files\Stata9\wsestata.exe" /b do bigjob
```

This tells Stata to execute the commands in `bigjob.do`, suppress all screen output, and route the output to `bigjob.log` in the same directory. If you desire a SMCL log file rather than an ASCII file, specify `/s` rather than `/b`.

If the do-file loads datasets that require more than the default amount of memory (10 megabytes for Stata/SE, 1 megabyte for Intercooled Stata), you will need to allocate this to Stata when you start it. Typing

```
C:\data> "C:\Program Files\Stata9\wsestata.exe" /m15 /b do bigjob
```

will run `bigjob.do` with 15 MB of memory. Alternatively, `bigjob.do` can try to increase the memory allocated to Stata while it is running with the `set memory` command; see [GSW] **C. Setting the size of memory**.

While the do-file is executing, the Stata icon will appear on the taskbar together with a rough percentage of how much of `bigjob.do` Stata has executed. (Note that Stata calculates this percentage based on the number of characters in `bigjob.do`, so the percentage may not accurately reflect the amount of time left for the job to complete.)

If you click the icon on the taskbar, Stata will display a box asking if you want to cancel the batch job.

Once the do-file is complete, Stata will flash the icon on the taskbar on and off. You can then click the icon to close Stata. If you wish for Stata to automatically exit after running the batch do-file, use `/e` rather than `/b`.

Note that you do not have to run large do-files in batch mode. Any do-file that you run in batch mode can also be run interactively. Simply start Stata, type `log using` *filename*, and type `do` *filename*. You can then watch the do-file run, or you can minimize Stata while the do-file is running.

A.10 Launching by double-clicking on a .dta dataset

The first time that you start Stata for Windows, Stata registers with Windows the actions to perform when you double-click on certain types of files. For example, if you double-click on a Stata dataset (`.dta` file), Stata will start and attempt to `use` the file. If you are using Stata/SE, the default amount of memory allocated to data will be 10 megabytes; for Intercooled Stata, the default amount of memory allocated to data will be 1 megabyte.

To change the default amount of memory, you may either reset the permanent default memory allocation by typing

```
. set memory #m, permanently
```

or modify the **Properties** for `.dta` datasets:

1. Double-click on **My Computer**. Select **Tools > Folder Options...** (in some versions of Windows, you may instead select **View > Options**).

2. Click on the **File Types** tab. Scroll through the list until you find **Stata Dataset**. Click once on it, and then click the **Advanced** button (in some versions of Windows, you may instead click the **Edit...** button).

3. You will see an **Edit File Type** dialog. In the Actions section, click on **open**, and then click the **Edit...** button.

4. In the resulting dialog box, there will be an edit field containing something like

   `"C:\Program Files\Stata9\wsestata.exe" use "%1"`

   This is the DOS-like command to start Stata with a dataset. Notice its similarities to the **Target** of a regular Properties Sheet. When you double-click on a Stata dataset, Windows executes this command line, replacing the %1 with the path and filename of the file on which you double-clicked.

   You can add a /m option to specify the amount of memory to allocate when double-clicking on a .dta file. If you want Stata to start with 20 megabytes of memory, add a /m20 option before the use command:

   `"C:\Program Files\Stata9\wsestata.exe" /m20 use "%1"`

5. Once you have set /m as you wish, click **OK**, and then click **OK** from the **Edit File Type** dialog. Then, click **OK** from the **Folder Options** dialog.

Note that the /m option set here is for all .dta datasets, not a specific one. Also note that Small Stata users cannot vary the amount of memory Stata allocates to itself.

What if you want Stata to do other things as well? What if you have written a Stata program called myuse (see [U] **17 Ado-files**) that contains a list of commands that you want Stata to execute when it loads a dataset on which you double-click?

The trick is to change the command to be executed from

   `"C:\Program Files\Stata9\wsestata.exe" use "%1"`

to

   `"C:\Program Files\Stata9\wsestata.exe" myuse "%1"`

myuse.ado might read

```
───────────────────────────────────────────────── top of myuse.ado ─────────
 . . .
 if "`1'"!="" {
         use "`1'"
 }
 . . .
────────────────────────────────────────────────── end of myuse.ado ─────────
```

where . . . represents the other commands in myuse.ado file. That is, we just add these three lines to the file at the point where we wish to have the dataset loaded.

A.11   Launching by double-clicking on a do-file

Double-clicking on a do-file works just like double-clicking on a dataset. When you double-click on a do-file, Stata launches in interactive mode, executes the do-file, and then issues a prompt so that you can continue the session or exit.

You set the memory for a do-file in the same manner that you do for a `.dta` dataset. This time when you search for the file type, you click on **Stata do-file** rather than **Stata Dataset**.

The edit field containing the command will say do rather than `use`:

```
"C:\Program Files\Stata9\wsestata.exe" do "%1"
```

Note that if you add a `/m` option here, it is for all invocations by double-clicking on a do-file, not a specific one, and it is unrelated to whatever value you may have set for double-clicking on a `.dta` dataset. Also note that Small Stata users cannot vary the amount of memory Stata allocates to itself.

If you right-click on a do-file, you can select **Open** or **Edit**. **Open** does the same as double-clicking on the do-file. **Edit** starts Stata, opens the Do-file Editor, and loads the do-file. You can edit the action for **Edit** in the same way as described above for the **Open** action if you wish to use an editor other than Stata's own.

A.12   Launching by double-clicking on a .gph graph file

When you double-click on a Stata `.gph` file, Stata launches and displays the graph. The current directory is the directory containing the graph. Stata remains running. Type `exit` or click the close box to close when you are through with the session.

When you right-click on a Stata `.gph` file, you can select **Print** or **Print to default**. If you select **Print**, Stata launches, displays the graph, and opens a print dialog. If you select **Print to default**, Stata launches and prints the graph to the default printer, bypassing the print dialog. When the graph is finished printing (or when you cancel the print), Stata exits.

A.13   Running simultaneous Stata sessions

Each time you double-click on the Stata icon or launch Stata in any other way, you invoke a new instance of Stata, so if you want to run multiple Stata sessions simultaneously, you may. The title bar of each new Stata that is invoked will reflect its instance number.

B Troubleshooting starting and exiting Stata

Contents

B.1   If Stata does not start

You tried to start Stata and it refused; Stata or your operating system presented a message explaining that something is wrong. Here are the possibilities:

Cannot find license file

This message means just what it says; nothing is too seriously wrong. Stata simply could not find what it is looking for. The two most common reasons for this are that you did not complete the installation process or Stata is not installed where it should be.

Did you insert the codes printed on your paper license to unlock Stata? If not, go back and complete the initialization procedure.

Assuming that you did unlock Stata, Stata is merely mislocated, or the location has not been filled in.

Error opening or reading the file

Something is distinctly wrong and for purely technical reasons. Stata found the file that it was looking for, but either the operating system refused to let Stata open it or there was an I/O error. About the only way this could happen would be a hard-disk error. Stata technical support will be able to help you diagnose the problem; see [U] **3.9 Technical support**.

License not applicable

Stata has determined that you have a valid Stata license, but does not apply to the version of Stata that you are trying to run. You would get this message if, for example, you tried to run Stata for Windows using a Stata for Macintosh license.

The most common reason for this message is that you have a license for Intercooled Stata, but you are trying to run Stata/SE, or you have a license for Small Stata, but you are trying to run Intercooled Stata or Stata/SE. If this is the case, reinstall Stata, making sure to choose the appropriate flavor.

Other messages

The other messages indicate that Stata thinks you are attempting to do something that you are not licensed to do. Most commonly, you are attempting to run Stata over a network when you do not have a network license, but there are a host of other alternatives. There are two possibilities: either you really are attempting to do something that you are not licensed to do, or Stata is wrong. In either case, you are going to have to call us. Your license can be upgraded, or, if Stata is wrong, we can provide codes over the telephone to make Stata stop thinking that you are violating the license.

B.2   verinst problems

Once Stata is running, you can type `verinst` to check if it is correctly installed. If the installation is correct, you will see something like

```
. verinst
You are running Stata/SE 9.0 for Windows.
Stata is correctly installed.
You can type exit to exit Stata.
```

If, however, there is a problem, `verinst` will report it.

In most cases, `verinst` itself tells you what is wrong and how to fix it. There is one exception:

```
. verinst
unrecognized command
r(199);
```

This most likely means that Stata is not correctly installed. If you need to install Stata in a nonstandard way, you can learn exactly how Stata works by reading [GSW] **A. More on starting and exiting Stata**. Otherwise, reinstall Stata following the instructions in chapter 1.

B.3   Troubleshooting

Crashes are called Application Faults in Windows. The dreaded Application Fault is not always the application's fault. It can be caused by configuration problems, bugs in device drivers, memory conflicts, and even hardware problems.

If you experience an Application Fault, first make sure that you are running the most-current version of Stata (see chapter 19 for information on updating). If the problem still exists, look at the Frequently Asked Questions (FAQ) for Windows in the user-support section of the Stata web site *http://www.stata.com*. You may find the answer to the problem there. If not, we can help, but you must give us as much information as possible.

Reboot your computer, restart Stata, and try to reproduce the fault, writing down everything you do before the fault occurs. We will want that information, along with the contents of your CONFIG.SYS, AUTOEXEC.BAT, SYSTEM.INI, and WIN.INI files. If you cannot email them to us, at least print them so you can look at them when you call us.

If Stata used to work on your computer but suddenly stopped working, try to remember any hardware or software that you have recently installed.

In addition, give us as much information about your computer as possible. Is it a Pentium? A Celeron? What version of Windows are you running? What brand is your computer? What kind of mouse do you have? What kind of video card?

Finally, we need your Stata serial number and the date your version of Stata was "born". Include them if you email, and know them if you call. You can obtain them by typing `about` in Stata's Command window. `about` lets us know everything about your copy of Stata, including the version and the date it was produced.

C   Setting the size of memory

Contents

C.1   Memory-size considerations

Stata works with a copy of the dataset that it loads into memory.

By default, Small Stata allocates about 300K to Stata's data areas, and you cannot change it.

By default, Intercooled Stata allocates 1 megabyte to Stata's data areas, and you can change it.

By default, Stata/SE allocates 10 megabytes to Stata's data areas, and you can change it.

You can even change the allocation to be larger than the physical amount of memory on your computer because Windows provides virtual memory.

Virtual memory is slow but adequate in rare cases when you have a dataset that is too large to load into real memory. If you use large datasets frequently, we recommend that you add more memory to your computer. See [GSW] **C.4 Virtual memory and speed considerations**.

You can change the allocation when you start Stata, as was discussed in [GSW] **A.6 Specifying the amount of memory allocated**.

In addition, you can change the total amount of memory allocated while Stata is running and optionally make that setting the default to be used by future invocations of Stata. That is the topic of this chapter.

It does not much matter which method you use. Being able to change the total on the fly is convenient, but even if you cannot do this, it just means that you must specify it ahead of time, and if later you need more, you must exit Stata and reinvoke it with the larger total.

Do not specify more memory than you actually need. Doing so does not benefit Stata; in fact, it may slow Stata down!

C.2   Setting the size on the fly

Assume that you have changed nothing about how Stata starts, so you have the default amount of memory (10 megabytes for Stata/SE; 1 megabyte for Intercooled Stata) allocated to Stata's data areas. You are working with a large dataset and now wish to increase memory to 32 megabytes. You can type

```
. set memory 32m
```

and, if your operating system can provide the memory to Stata, Stata will work with the new total. Later in the session, if you want to release that memory and work with only 2 megabytes, you could type

```
. set memory 2m
```

173

There is only one restriction on the `set memory` command: whenever you change the total, there cannot be any data already in memory. If you have a dataset in memory, you save it, clear memory, reset the total, and then use it again. We are getting ahead of ourselves, but you might type

```
. save mydata, replace
file mydata.dta saved
. drop _all
. set memory 32m
...
. use mydata
```

When you request the new allocation, your operating system might refuse to provide it:

```
. set memory 512m
op. sys. refuses to provide memory
r(909);
```

If that happens, you are going to have to take the matter up with your operating system. In the above example, Stata asked for 512 megabytes, and the operating system said no.

C.3   The memory command

`memory` helps you figure out whether you have sufficient memory to do something. If you are using Stata/SE, you could type `memory`:

```
. memory
```

	bytes	
Details of **set memory** usage		
overhead (pointers)	114,136	10.88%
data	913,088	87.08%
data + overhead	1,027,224	97.96%
free	21,344	2.04%
Total allocated	1,048,568	100.00%
Other memory usage		
set maxvar usage	1,816,666	
set matsize usage	1,315,200	
programs, saved results, etc.	772	
Total	3,132,638	
Grand total	4,181,206	

You can type `memory` in Intercooled Stata, too, but the output will vary slightly from Stata/SE; see [D] **memory**. 21,344 bytes free is not much. You might increase the amount of memory allocated to Stata's data areas by specifying `set memory 2m`.

```
. save nlswork
. set memory 2m
...
. use nlswork
(NLS Women 14-26 in 1968)
```

```
. memory
```

	bytes	
Details of **set memory** usage		
overhead (pointers)	114,136	5.44%
data	913,088	43.54%
data + overhead	1,027,224	48.98%
free	1,069,920	51.02%
Total allocated	2,097,144	100.00%
Other memory usage		
set maxvar usage	1,816,666	
set matsize usage	1,315,200	
programs, saved results, etc.	773	
Total	3,132,639	
Grand total	5,229,783	

Over 1 megabyte is now free; that's better. See [D] **memory** for more information.

C.4   Virtual memory and speed considerations

When you open (use) a dataset in Stata for Windows, Stata loads the entire dataset into memory.

When you use more memory than is physically available on your computer, Stata slows down. If you are using only a little more memory than is on your computer, performance is probably not too bad. On the other hand, when you are using a lot more memory than is on your computer, performance will be noticeably affected. In these cases, we recommend that you

```
. set virtual on
```

Virtual memory systems exploit locality of reference, which means that keeping objects closer together allows virtual memory systems to run faster. set virtual controls whether Stata should perform extra work to arrange its memory to keep objects close together. By default, virtual is set off. set virtual can only be specified if you are using Stata/SE or Intercooled Stata for Windows.

In general, you want to leave set virtual set to the default of off so that Stata will run faster.

When you set virtual on, you are asking Stata to arrange its memory so that objects are kept closer together. This requires Stata to do a substantial amount of work. We recommend setting virtual on only when the amount of memory in use drastically exceeds what is physically available. In these cases, setting virtual on will help, but keep in mind that performance will still be slow. See [U] **6.5 Virtual memory and speed considerations**. If you are using virtual memory frequently, you should consider adding memory to your computer.

Notes

D More on Stata for Windows

Contents

D.1   Using Stata datasets and graphs created on other platforms

Stata will open any Stata `.dta` dataset or `.gph` graph file regardless of the platform on which it was created, even if it was a Macintosh or Unix system. In addition, Stata for Macintosh and Stata for Unix users can use any files that you create. If you transfer a Stata file using file transfer protocol (FTP), just remember to transfer using binary mode rather than ASCII.

If you need to send a dataset to a colleague who is using Stata 7.0, first recommend that they upgrade to the latest version of Stata. Then save your dataset with the `saveold` command, so your coleague will be able to read it even if they do not upgrade. For example,

```
saveold "C:\data\flood2.dta"
```

D.2   Exporting a Stata graph into another document

The easiest way to export a Stata graph into another application is via the Clipboard.

Create your graph. To copy it to the Clipboard, right-click on the Graph window, and select **Copy**. Stata will copy the graph as an Enhanced Metafile (EMF); this ensures that the receiving application obtains it in the highest resolution possible. If the receiving application does not understand the EMF format, select **Prefs > Graph Preferences**, click on the **Clipboard** tab, and click the **Windows Metafile (WMF)** radio button to copy graphs in the Windows Metafile (WMF) format.

A metafile contains the commands necessary to redraw the graph. That is, a metafile is a collection of lines, points, text, and color information. Metafiles, therefore, can be edited in a structured drawing program.

After you have copied the graph to the Clipboard, switch to the application into which you wish to import the graph and paste it. In most applications, this is accomplished by selecting **Edit > Paste**. Consult the documentation for your particular application for more details.

Stata also gives you the ability to create a file containing the metafile. You could later import this file into another application. To save a graph as a metafile, right-click on the Graph window, and select **Save Graph...**. In the **Save** dialog, choose **Windows Metafile** as the file type.

Stata can save a graph as an Encapsulated PostScript (EPS) files for importing into desktop-publishing applications. To save a graph as an EPS file, right-click on the Graph window, and select **Save Graph...**. In the **Save** dialog, choose **EPS** as the file type. If you wish to include a preview of the graph so that it may be viewed in your desktop-publishing application, choose **EPS with TIFF preview**. Choosing the preview option does not affect how the graph is printed. See the *Stata Graphics Reference Manual* for more information.

D.3   Saving the contents of the Review window as a do-file

To save the contents of the Review window as a do-file, right-click on the Review window and select **Save Review Contents...** Stata will save the commands in the file you specify in the **Save** dialog. You can edit the do-file in Stata's Do-file Editor (see chapter 14 earlier in this book), or in a text editor, such as Notepad, if you wish. If you use a word processor to edit the file, make sure to save the file as text. Also, be aware that some editors, such as Notepad, will always try to append a .txt extension to the filename. Even though you may have typed myfile.do in the **Save as** dialog box, Notepad will save myfile.do.txt. To force such an editor to use the extension you want, enclose the filename in double-quotes; type "myfile.do" rather than myfile.do.

You can also copy the contents of the Review window to the Clipboard by right-clicking on the Review window and selecting **Copy Review Contents to the Clipboard**. You can then paste the contents into Stata's Do-file Editor or any other text editor.

For more information on do-files, see [U] **16 Do-files**.

D.4   Installing Stata for Windows on a network drive

You will need a network license before you can install Stata on a network drive. You can install Stata from the server, or if you have the appropriate privileges, you can install Stata directly to the network drive.

Follow the instructions for installing Stata in chapter 1. Depending on where you wish to provide network access to Stata from, it may be necessary to change the installation directory at step 5 of the installation instructions.

Once Stata is installed, run it to initialize the license. Mount the network drive that Stata is installed on from a workstation. Right-click on the Desktop or on the Windows **Start** menu, and select **New > Shortcut**. Type the path for the Stata executable into the edit field, or click **Browse...** to locate it. Enter Stata for the name of the shortcut.

Once a shortcut for Stata has been created, right-click on it, and select **Properties**. Set the default working directory for Stata by changing the **Start in:** field to a local drive to which users have write access. This is where Stata will store datasets, graphs, and other Stata-related files. If the workstation will be used by more than one user, consider changing the **Start in:** edit field to the environment variable %HOMEDRIVE%%HOMEPATH%. This will set the default working directory to each user's home directory.

D.5   Changing a Stata for Windows license

If you've already installed Stata and your license needs to be changed,

1. You will need the *License and Authorization Key*.

2. Open the Stata directory from the hard drive.

3. Right-click on the STATA.LIC file, and select **Rename**.

4. Change the name of the file to STATA.LIC.old.

5. Run Stata, which will then prompt you for your license information and recreate the STATA.LIC file.

6. Run Stata as an unprivileged user to ensure that it runs correctly. If Stata runs for the privileged user but does not run for the unprivileged user, check the permissions on the STATA.LIC file.

7. Once Stata runs properly with the new license information, delete the file STATA.LIC.old.